Apprai___ and
Selection of Projects

Appraisal and Selection of Projects
A Multi-faceted Approach

Utpal K. Ghosh

CRC Press
Taylor & Francis Group
Boca Raton London New York

CRC Press is an imprint of the
Taylor & Francis Group, an **informa** business

First edition published 2022
by CRC Press
2 Park Square, Milton Park, Abingdon, Oxon, OX14 4RN

and by CRC Press
6000 Broken Sound Parkway NW, Suite 300, Boca Raton, FL 33487-2742

British Library Cataloguing-in-Publication Data
A catalogue record for this book is available from the British Library

ISBN: 978-1-032-04228-2 (hbk)
ISBN: 978-1-032-04231-2 (pbk)
ISBN: 978-1-003-19103-2 (ebk)

Typeset in Times
by codeMantra

To
Manjula,
Supriti,
and
Indranil

Contents

Preface

Project appraisal is the process of assessing in a systematic way, the viability of a project at its initial 'idea' stage when information is minimal, but the decision, whether or not to go ahead with the project, is important. It is in the backdrop of this situation that this book is immensely relevant.

This book deals with the principles and practice of project appraisal techniques and thus should be of deep interest and value not only to the practicing project managers in the office, but also to entrepreneurs, teachers and students in academia, advisors and decision makers in industry and public administration.

Normally, books on appraisal of projects available in the market deal primarily with the financial and economic aspects. This book deals with the above issues as well as other issues such as:

- Market analysis;
- Technical analysis;
- Environmental appraisal;
- Life cycle costing;
- SWOT analysis.

Thus, it deals with multiple issues in the area of appraisal and selection of projects.

The text can be broadly divided into three parts. The first part comprises Chapters 1 and 2 which deal with the basics of the subject, viz., the characteristics and success of a project and its appraisal methods. Chapters 3 through 8 primarily form the second part and deal with the analysis of the project proposals from different perspectives for ascertaining their viabilities, including a discussion on uncertainty and risk analysis. The third part comprises the concluding Chapters 9 and 10 which deal with the life cycle costing and SWOT analysis, which form the two important tools or techniques that can be used for project appraisal.

The material covered in the text has been drawn from the vast pool of accumulated knowledge and experience of distinguished analysts gained through studies carried out in different countries, and supplemented by the author's own experience. As far as possible, references of the published literature have been mentioned at the end of each chapter. The author thankfully acknowledges his indebtedness to these writers. However, if ideas of earlier writers have appeared in this book without appropriate acknowledgement, it is quite unintentional and the author would like to extend apologies. If such instances are brought to the notice of the author, the same will be gratefully acknowledged in the subsequent edition of this book.

In writing this book, the author has also gained enormously from interactions with a large number of individuals. Many a times small points raised in the discussions have led to major change in the text or the inclusion of an additional topic. It is practically not possible to list such individual names. However, the author gratefully acknowledges his debt to each of them. Special thanks are due to Ranjit Kumar Banerjee, formerly of Asian Development Bank, and Sudhansu Kumar

Bhattacharjee, formerly of Braithwaite, Burn & Jessop Construction Co Ltd, for their support throughout the preparation of the manuscript and valuable suggestions. The author thanks Rashmi Rakshit Mallick for drawing the figures for this book. Thanks are also due to Tilokesh Mallick for the long hours he spent for helping the author in surfing the internet and for keying in the bulk of the text in the computer.

And last, but by no means the least, the author is grateful to his wife Manjula, daughter-in-law Supriti and son Indranil for their fullest co-operation, encouragement and support in writing this book.

Author

Uptal K. Ghosh received his civil engineering degree in 1954 from Bengal Engineering College, Shibpur, Calcutta University (currently Indian Institute of Engineering, Science and Technology). He worked, among others, with Freeman Fox and Partners, London; Sir William Arrol & Co Ltd, Glasgow; and Braithwaite, Burn & Jessop Construction Co Ltd (BBJ), Kolkata. Subsequently, he set up his own practice as consulting engineer.

During his long career, he participated in the planning, design, fabrication, erection and overall management of a wide variety of projects such as bridges and industrial structures, which included new construction as well as repair and rehabilitation work. Countries covered by his work experience include the United Kingdom, New Zealand, Malaysia, Indonesia, Singapore, Nepal and India.

He has published a number of papers, and is the author of three books entitled *Design and Construction of Steel Bridges*, *Repair and Rehabilitation of Steel Bridges* and *Design of Welded Steel Structures*.

He is a chartered engineer and a fellow of the Institution of Engineers (India), a member of the Institution of Civil Engineers (U.K.) and a Member of the Institution of Structural Engineers (U.K.).

1 Characteristics of a Project

1.1 INTRODUCTION

Change is the essence of human development, and project may be considered as the principal vehicle for this change. Both change and project are associated with human race for a long time. Perhaps the first project of humankind was to build his own habitat – a simple house, and shift there from the cave to live in. Similarly, placing a few branches of trees across a small river to cross it was also an example of an early project executed by man. Advancing further into many centuries of human advancement, we can cite examples of magnificent and formidable projects such as the Great Pyramid of Giza in Egypt, Mayan temples in Central America, Mohenjo-Daro structures in ancient India, Acropolis of Greece and the Great Wall of China, to name only a few. These ancient projects were probably executed without much control on expenses or completion time, and were consequently not built frequently, perhaps no more than once in a generation.

In contrast, in our present-day life, projects are regular features and have become a part of modern human activities. Furthermore, we often come across a number of projects being implemented simultaneously. In addition, these projects are of much wider varieties than were observed ever before. This is primarily because project is a change-creating mechanism, and we have all become aware of its power to use our resources more efficiently to bring about this change. The mechanism can be used by a large customer-oriented organization to deploy its resources efficiently in order to respond quickly to the needs of the target customers.

Eiffel tower in Paris, Taj Mahal in Agra (India), Sydney Opera House, Sutong Bridge (1,088-m-long main span) at Yangtze River Estuary in China, Burj Al-Arab hotel (321 m high) in Dubai etc. are examples of modern projects having comparatively shorter life spans. Illustrations of mega projects are Channel Tunnel between England and France (50.5 km long), China's proposed project to connect the Yangtze, Hai, Huai and Yellow rivers (expected duration: 40 years), Greece's Egnatia Highway (670 km long), etc. Similarly, the mechanism of project can be used to benefit small groups of individual homes to enhance their quality of life more efficiently. Grameen Bank concept for microfinancing self-help groups of needy rural women in Bangladesh is a noteworthy example of a mega project for providing social benefit in a developing country.

1.2 WHAT IS A PROJECT?

One can define a project as an activity involving two or more parties working together in an enterprise involving a number of activities. In other words, we can also define project as a collection of interrelated tasks to be completed within a fixed period and limited cost.

1.2.1 DEFINITION

Over the years, many professional bodies and individuals have, in many different ways, answered the question, 'What is a project?'; however, the definition given in ISO 21500:2012 'Guide to Project Management' is perhaps the most appropriate one. It defines project as

> a unique set of processes consisting of co-ordinated and controlled activities with start and finish dates, undertaken to achieve an objective. Achievement of the project objective requires deliverables conforming to specific requirements, including multiple constraints such as time, cost and resources.

It is apparent from this definition that a project is essentially a vehicle for change and there is a fundamental difference between a project and an ongoing organization. While the former is fundamentally a process of change, the latter carries on with the ongoing operations in a continuous, repetitive and routine manner, often following a well-defined route. Any small improvement (change) in the latter case is performed as a part of day-to-day work without affecting the overhead structure.

1.2.2 DISTINCTIVE FEATURES

Irrespective of their size and nature, all projects are marked by a few common distinctive features. Thus, every project is:

- A mechanism for change with the well-defined goal;
- A cost-oriented venture;
- Time-bound;
- Temporary in nature;
- Unique;
- Resource-dependent;
- People-oriented;
- Subject to uncertainty and risk.

These aspects are briefly discussed in the following paragraphs.

1.2.2.1 Change

Project is a vehicle for change. It provides the means by which an individual, an organization or a country can change from one stable state to another distinctly different stable state, from which the desired operation can be carried out, ideally, in a better way. The change in the operation may be quite significant or trifling depending on the expanse or outreach of the project. Project is a linked set of activities for producing an outcome to achieve a well-defined goal. Examples are constructing a new bridge across a river, launching a new magazine, expanding a business or enterprise and campaigning for election in a local body.

1.2.2.2   Cost

Every project needs investment by way of people's time, equipment, computer hardware, computer software licenses, materials, marketing expenses, etc. Thus, there must be a well-defined budget before the project is conceived. This budget needs to be monitored, and every effort should be made to ensure that the project is completed within the agreed budget.

1.2.2.3   Time

The duration for completion may vary from project to project, depending on its size and involvement. However, each project must have a limited and well-defined period of completion time to produce an outcome of certain standard.

1.2.2.4   Temporary in Nature

Every project is a one-off event. It may be installing a new system or software in the office, building a new hospital, replacing machine in a workshop or rehabilitating an existing bridge. In every case, the project is intentionally temporary; it will come and go, leaving behind an intended outcome. Any delay in its completion would not only delay the outcome of the project but also affect the release of the assigned resources for use in other project(s).

1.2.2.5   Uniqueness

Each project is a unique enterprise. No two projects are identical, even though the associated activities may be similar and their outcomes may be the same. As for example, two similar bridges over two different rivers may look alike, but they will have different site conditions, different design norms to comply with, different foundations, different contractors and so on.

1.2.2.6   Resource Dependence

Every project – big or small – is always resource-dependent. Professionals, experts as well as materials and equipment are required for successful execution of any project, right from the conception stage till the project is completed. However, all the resources may not be required simultaneously for the entire duration of the project. Thus, resources for, say, one particular activity may be required only after completion of another preceding activity. Also, it may be that two or more activities may run in parallel, in which case resources for these particular activities will be needed simultaneously. Therefore, to avoid wastage, ideally, resources for each activity should be made available only when needed and taken away when the need is over.

1.2.2.7   Focus on People

All projects are people-oriented and consequently associated with various groups of persons, commonly known as *stakeholders*. Stakeholders are intimately involved in a project, though their interests or priorities may not be similar and may even be in opposition. This aspect needs to be specially noted. For example, one of the stakeholders is the sponsor of a project. This may be an individual or an organization where priority could be to achieve a maximum profit from the project with a minimum cost. On the contrary, the priority of a government or nonprofit private organization would be to spread the benefit to the maximum number of people (users). Next, a project will need skilled people to conceive, plan and execute it to successful completion. The stakeholders in this group need to be specialists in the respective nature of the project, viz., a railway bridge project, a cancer hospital project or developing new software, etc., as the case may be. The priority of this group is to ensure that the outcome becomes beneficial to the users, in spite of unavoidable uncertainty and possible risks. The third and the most important group of stakeholders is the end users or beneficiaries of the project who will be concerned primarily with the usefulness of the outcome of the project. Depending on the nature of the project, other stakeholders include the politicians, regulatory authorities, environmentalists and social activists, media.

One other aspect regarding 'cross-linked' stakeholders needs consideration in this context. A project may need diverse types of expertise from several organizations, each of whom, in turn, may be engaged by several project authorities for its distinctive expertise. The competing demands of these diverse projects may initiate conflicts in the deployment of resources of these expert organizations, creating problems to the customers (the diverse project authorities) in not receiving the expert inputs in time, thereby exposing the project to the risk of slippage in time schedule.

1.2.2.8   Uncertainty and Risk

As already discussed, every project is linked with change, and change always entails a certain degree of uncertainty and risk. Apart from this inbuilt uncertainty, one has to be alert from a number of extraneous uncertainties also. Natural environment has always been uncertain. Earthquakes, floods and hurricanes are examples of such uncertainties. Of late, political, social, economic and technological environments around us have undergone enormous change and are still continuing to do so. As a result, unprecedented instability is being experienced in all spheres of our life, leading to uncertainty and risk. Thus, it is rather impossible to make any project completely free from uncertainty and risk.

1.2.3   Basic Features

Out of the distinctive features discussed in the previous section, every project has three basic features of topmost importance, which must be monitored throughout the duration of the project. These are:

- Intended outcome, i.e., 'will the outcome perform its intended task?'
- Time, i.e., 'how long will it take to complete the project?'
- Cost, i.e., 'what will be the expense?'

All these features are not only important by themselves, but are at the same time interrelated. In many projects, one or more of the basic features often need adjustment to meet changed priorities. Thus, during the progress of the project, it may be found that the project is running behind schedule. This deficiency may be made up by inducting extra resources (e.g., labor, equipment) to expedite the critical activities. This would, of course, change the estimated project cost. Alternatively, in order to complete the project within the specified time without incurring extra expense, the intended outcome may have to be compromised and some subsequent project activities modified to keep to the intended time schedule. It also happens that some new activities are added during the progress of the project to achieve additional benefits. In such a case, both time and cost are likely to be increased. Such adjustments of the project profile are quite normal, and it is therefore necessary to closely monitor these basic features of every project to achieve success at the end. Added to these basic features, involvement of the project organization structure is the other internal feature, which largely contributes to the success of a project. Figure 1.1 depicts a model of these four interrelated basic features of a project.

Furthermore, as discussed earlier, there are other external issues that are also associated with the smooth running of a project. These may be collectively termed project surroundings and include issues like timely supply of resources, uncertainty/risk and safety as well as observations/opinions of other stakeholders such as end users, technical experts, consultants, environmentalists and social activists.

1.3 PROJECT LIFE CYCLE

Every project has a life cycle consisting of a number of distinct, value-added and interlinked stages through which it has to proceed in a logical sequence from start to completion. It is a dynamic process in which its status commences from 'idea (of change for improvement) stage' to 'appraisal stage', followed by actual implementation of the project and culminating in its termination and delivery of the benefits.

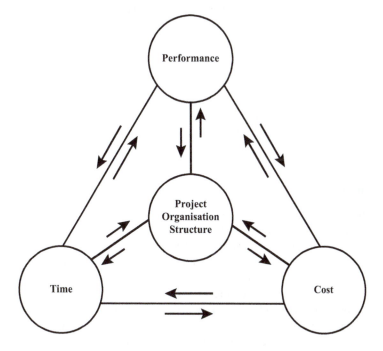

FIGURE 1.1   Model showing the four interrelated basic features of a project.

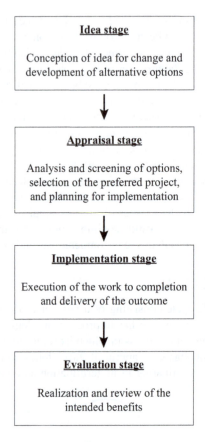

FIGURE 1.2 Typical life cycle of a project.

Normally, a post-project evaluation of realization of the intended benefits is also undertaken. A typical life cycle of a project showing the above stages is shown in Figure 1.2. The salient features of these stages are briefly discussed in the next sections.

1.3.1 IDEA STAGE

This is the first stage of a project when a need for change, i.e., a *prima facie* case for a development idea (change), is conceived, and a series of alternative options for achieving the objective of the development idea are proposed. For example, recurrent traffic congestion may be noticed in a particular road for a considerable period of time. A few alternative solutions, such as widening of the road and by introducing additional traffic lanes or constructing a bypass or a flyover, are offered. Similarly, in an increasing market demand, a special type of chemical product may be noticed and a few alternative schemes are proposed to meet the demand. It is implied that technical feasibility study, estimates of cost and time and other related aspects for successful completion of the options offered are to be examined in a rough but reasonably comprehensive manner before these are put forward for a detailed analysis and screening in the next stage.

1.3.2 APPRAISAL STAGE

In this stage, the various embryonic options for achieving the objective are analyzed in detail in respect of different aspects.

An important advantage of having a detailed analysis of the issues during appraisal stage is that a relatively small amount of money is involved in learning the success potential of the alternative options and also the uncertainty and risks involved before committing to spend large sums of money in implementing the project. However, awareness of the utility of 'appraisal process' among the stakeholders is sadly still relatively low.

1.3.3 IMPLEMENTATION STAGE

Implementation stage is the most critical stage in the project cycle, when major financial investment takes place. Typically, in terms of duration, this stage is the longest one. Also, the levels of activities reach their peak in this stage. It involves implementation of the diverse tasks evolved during the appraisal stage and finally delivering the outcome. Simultaneously, while these activities are underway, a series of management processes are taken up to monitor and control various aspects of the project. Some of these aspects are:

- Quality assurance;
- Time;
- Financial and other resources;
- Uncertainty and risk;
- Safety.

Once the deliverables have been physically produced to acceptable standards and approved, the project is deemed to have been completed and the outcome is then handed over. This stage also includes releasing the project resources, terminating the input contracts, handing over the relevant project records to the authorities, and finally notifying the stakeholders and interested parties.

1.3.4 BENEFITS DELIVERY STAGE: REALIZATION AND REVIEW OF THE BENEFITS

Once a project is complete, the benefits are obtained. The production starts, direct and indirect employments are created, and other demands are met.

In this stage, a review is undertaken to find out the level of success of the project, i.e., whether the benefits have been delivered as per expected standard and that too, within the stipulated time and financial limit. The findings of both achievements and failures are recorded as a post-implementation review, so that lessons learned from various activities can be utilized for appraisal of future projects.

BIBLIOGRAPHY

Baguley, P., 2008, *Project Management*, Chartered Management Institute/Hodder Education, London.
Batchelor, M., 2010, *Project Management Secrets*, Harper Collins, London.
Roberts, P., 2011, *Effective Project Management*, Kogan Page Ltd., London.
Smith, N.J., 1996, *Engineering Project Management*, Blackwell Science Ltd., London.
Westland, J., 2007, *The Project Management Life Cycle*, Kogan Page Ltd., London.

2 Success and Appraisal of a Project

2.1 INTRODUCTION

Basically, project is a vehicle for achieving certain specific objective, and once the objective has been fully realized, the project may be considered as completed and successful. However, as has been discussed in Chapter 1, the journey is not easy. Indeed, it is quite painstaking and fraught with risks of failure. Therefore, before dwelling on the topic of success of a project, the first task should be to understand the nature and range of success and ensure that the possible causes of failure are taken care of at the appraisal stage itself.

2.2 RANGE OF SUCCESS

As has been discussed in Chapter 1, every project has three basic features, viz., performance (specification), time and cost, which are required to be in line with the initial plan. If all these features are satisfied, the project is considered as successful. On the contrary, when a project is completed, but none of the three basic features are satisfied, the project is considered to have failed. The project may also be terminated early or abandoned before completion due to a variety of reasons. In such a case, the project is also deemed to have failed. These instances depict two extreme results. In practice, however, a project may have achieved or nearly achieved two or even only one of the basic requirements. In such a case, the success or failure would be only partial and not absolute. Thus, there may be degrees of success or failure depending on the levels of achieving the required basic features.

The range of such success/failure is collated below according to a diminishing range of success:

- Absolute success:
  - Project completed as per scheduled time, cost and specification fully.
- Partial success:
  - Project completed and meets two out of the three basic features;
  - Project completed, but meets only one out of the three basic features.
- Failure:
  - Project completed, but does not satisfy any of the basic features;
  - Project terminated before completion.

2.3   CRITICAL SUCCESS FACTORS

Probability of success of a project depends basically on the levels of analysis of critical success factors. These are those elements in the project cycle which must be given special and continued attention from the very beginning to ensure that the project is successfully completed. These factors may be divided into two groups according to the periods when these should be taken care of, viz.,

- Pre-implementation period;
- Implementation period.

There are quite a number of these factors which should be considered to achieve the goal. The more important ones are discussed in the paragraphs that follow.

2.3.1   PRE-IMPLEMENTATION STAGE

The factors that should be considered prior to the start of implementation stage are as follows:

- Definition of the objective;
- Investigation and analysis;
- Assessment of uncertainties and risks;
- Estimate of time and cost;
- Planning.

2.3.1.1   Definition of the Objective

The first exercise is to define a clear objective of the project. Basically, this means the goal where the journey should end. The objective should be specific, i.e., quantifiable and the quality should be measurable. Consequently, the project management team should share a coherent understanding of the expected objective.

2.3.1.2   Investigation and Analysis

Quite often, analyses of the different project options for achieving the objective are not carried out properly for want of sufficient data. This shortcoming is due to various reasons, such as superficial field investigation, inaccurate estimation of time, cost and benefits, etc., and particularly unreasonable haste in completing the study. As a result, such projects suffer in time and/or cost overrun or fail to achieve the specified objective, or are even terminated during the implementation stage.

It is therefore imperative that appraisal of a project should be carried out in detail, with all sincerity, and should be given sufficient time to complete properly. This effort at the initial stage of a project is expected to bring about high performance in the implementation stage.

These aspects will be discussed in greater detail in chapters that follow.

2.3.1.3   Assessment of Uncertainties and Risks

Having properly defined the objective of a project, assessing the uncertainties and risks involved in realizing the objective is a necessary step for achieving success. One must understand that uncertainty and risk are inseparable from any project, no matter how hard one tries to plan to force it from these unwelcome aspects. However, efforts must be made to reduce or at least limit the influence of *foreseeable risks* in order to enhance the chances of success.

The points that are required to be addressed in this respect are as follows:

- Identification of any foreseeable risk;
- Reduction or elimination of such risks if possible;
- Decision whether or not to allow the risk to remain while proceeding with the plan.

It is imperative that these aspects are considered and closely reviewed at the very early stage of the project initiation, so that any future untoward event does not take the implementing team by surprise and thereby thwart the progress of the project. This topic will be discussed in detail in a later chapter.

2.3.1.4   Estimate of Time and Cost

Estimating time to execute the project activities and cost of resources for these must be considered in unison as these two are interrelated and interdependent. This topic has been discussed in Chapter 1.

2.3.1.5   Planning

Next, it is necessary to proceed, at a very early stage, with the preliminary design work and prepare a plan for implementation. This will enable the stakeholders to study the project in entirety prior to commitment to the costs of manufacturing the required resources or construction at site.

The plan should be prepared in sufficient detail in order to avoid unexpected obstacles, and should provide a complete picture of how the project will be implemented *via* a series of activities in sequence, describing the processes, the responsibilities of the concerned participants, etc., and finally deliverables committed in the objective. It is relevant to add here that every project comprises a number of interdependent activities. Thus, one particular activity may have to wait till completion of another *particular* preceding activity. For example, in a building project, the foundations are required to be built before the corresponding superstructure can be constructed.

It is now common to use computers in the planning and monitoring of projects. Some organizations use standard planning software, while some use their own tailor-made version. In choosing standard software, care must be taken to ensure that the software is appropriate for the project plan to be used and also compatible with the hardware, operating system, etc., of the computers available with the organization. In case the existing facilities need change/replacement, the additional costs for the same should be included in the cost estimate.

2.3.2   IMPLEMENTATION STAGE

The factors that should be considered during the implementation stage are as follows:

- Financial support;
- Human relationship;
- Delegation of authority;
- Response to changes in the plan;
- Contract terms;
- Communications system;
- Monitoring and control;
- Safety aspect.

2.3.2.1   Financial Support

Quite often, physical progress of projects has been affected due to delay in the release of funds (commensurate with the requirement) due to financial constraint of the investing agency. This is likely to not only delay the completion time, but also increase the prices of goods and services due to inflation. In order to keep up to the target of time and cost, it is therefore imperative that every effort is made to ensure timely release of funds to achieve the scheduled completion time and avoid cost over-run and decreased outcome level.

2.3.2.2   Human Relationship

Every project is a people-centric endeavor. Thus, while good planning and financial support are important, these cannot produce good results on their own. They need *groups of people* to bring the project to a successful end. Therefore, it is necessary to have a satisfactory human relationship in every project. The key groups are as follows:

- Project team and its leader;
- Providers of goods and services;
- End users;
- Others.

In the following sections, the roles of some of the key groups for a successful project outcome are briefly discussed.

2.3.2.2.1   Project Team and Its Leader

Ideally, the project should have a coherent team by deliberately inducting people with complementary role, committed to a common objective and having alongside an inspired and dedicated leader. The team members should be driven by a team spirit – a sense of shared identity – who are likely to work together well, guide and help each other irrespective of their ranks, and hold themselves collectively accountable for the performance of the team. This team spirit should be developed and sustained professionally. The team members should be mobilized *in time*, suitably ahead of the commencement of actual implementation to enable them to understand, review and plan their work and develop their system of communications in advance. The art and science of modern project management need to be inculcated into all the members of the team. An uneducated and untrained workforce is likely to run counter to delivering a successful project outcome. This aspect should be given due importance from the initial stage of the project.

Every project needs a leader, usually termed Project Manager. Project Manager is expected to possess leadership skills. He should be able to develop rapport with his teammates and motivate them. He should also respond positively to the suggestions made by others, and avoid unilateral imposition of decisions on the teammates. It is one of his primary roles to use his skills in persuasion in dealing with disputes among members of the team and search for areas of agreement towards a consensus-based solution. He has also to deal with different stakeholders such as clients, technical, financial, commercial, legal and environmental experts involved in the project. He should therefore have the ability to communicate with them positively and act as a shock-absorber in finding ways of resolving the conflicts between their very different interests. Furthermore, he is expected to possess contract management ability. He should be able to plan in detail, monitor and control the progress of the project, and also to anticipate problems and solve them by making on-the-spot decisions wherever possible. In short, he should be able to develop and manage a well-oiled operational 'team vehicle' in a scenario of constant change and drive it to its intended destination.

2.3.2.2.2 *Providers of Goods and Services*

More often than not, a project requires external resources such as goods and services to achieve its goal. These external stakeholders should be chosen with care, considering the time frame and cost aspect, as well as their quality of performance. For success of a project, 'participating type' of management should be established. These external contributors should be involved in the decision-making process from the very beginning. They should also be encouraged to train and brief their staff about the management setting of the project and its overall supporting environment, including other stakeholders. This would help in attaining the objective successfully.

2.3.2.3 Delegation of Authority

In order to avoid a delay in giving decisions for urgent problems that require prompt action, it is necessary to introduce a system of providing appropriate delegation of authority to key members involved in the project, particularly when the location of the project is remote from the headquarters. This would forestall avoidable crises, extra expenditure as well as erosion of confidence in the team. Therefore, care should be taken to ensure that the concerned members are capable and possess enough experience to make quick decisions on matters delegated to them.

2.3.2.4 Response to Changes in the Plan

In spite of all efforts, quite often, a project gets delayed during the implementation stage due to various reasons. Critical activities take longer time period, thereby affecting the overall completion schedule. Also, sometimes due to changed situation, the intended goal may have to be revised, entailing alteration to some of the activities. It is therefore necessary and advantageous to introduce a certain amount of in-built flexibility, particularly in the time period estimates of different activities. With such a planning, the obligatory changes may be accommodated without sacrificing the ultimate outcome of the project.

2.3.2.5 Contract Terms

Contract terms with the external stakeholders, viz., the providers of goods and services, should be fair and equitable, so that all the parties are motivated to try their best to achieve their objectives and help to the success of the project.

The terms and conditions in the contracts between the customers, suppliers, contractors, subcontractors, etc. should be unambiguous, and the rights and responsibilities of each party should be clearly allocated, to avoid, or at least minimize, the number of contractual disputes. This would help smooth running of the project immensely.

2.3.2.6 Communications System

For the success of a project, it is vital that a communications system be introduced from the early stage of the project. The system involves collecting different information regarding the project, and reaching it to the right people as a matter of routine for dissemination and further action if necessary. It *inter alia* includes formal project status report, any information that might affect the progress now or in the future leading to group discussions, and review future plans. Each information should be properly documented and retained for future reference.

2.3.2.7 Monitoring and Control

Once a project is launched, it becomes imperative to introduce a chain of systems to monitor and control the series of activities being undertaken by the project team. These systems primarily involve auditing the performance against the plan, ascertaining the causes of deviation, if any, finding an appropriate solution to the problem causing the deviation and taking corrective actions and changing plans of future activities in order to put the project back to its intended course.

The process also involves guarding against the forces that might affect the project now or in the future, and making suitable change now. Change may cost money. The organization should therefore estimate the cost of *not* removing the problem, and then take the decision after evaluating the pros and cons of the two alternatives. Some of the important aspects that should come under the purview of the monitoring and control are as follows:

- Time;
- Cost;
- Quality;
- Communications;
- Supply of resources;
- Release of funds;
- Uncertainty and risk.

Satisfactory monitoring and control of these aspects is essential for the success of any project, big or small.

2.3.2.8   Safety Aspect

Another aspect that warrants attention is safety during the project stage and also after its completion. Nowadays due to highly competitive environment with a tendency to speed up progress to meet the dead line, complacence and negligence to workmanship or even deficiency in the quality of material and inputs used are not uncommon, unless strict quality control regime is enforced. Thus, a project may be prone to safety hazards and accidents, not only during its execution stage, but also after the project is completed. This aspect is particularly important in industries like airlines, railways, mining and nuclear power, where safety aspect must be considered to be of paramount importance. Compromising in maintenance requirements due to the lack of funds is a common hazard in completed projects. This situation should also be brought into focus at the concept stage itself. Under no circumstances, safety aspect can be compromised for the sake of early completion of the project or budget constraint. The disasters of Chernobyl Nuclear Power Plant at Pripyat in Ukraine on 26 April 1986, or the toxic gas poisoning incident in Bhopal, India, on 2/3 December 1984, considered the world's worst industrial disaster consuming more than 10,000 lives, are not easy to forget.

2.3.3   Sponsor

As already discussed in Chapter 1, the sponsor of a project can be an individual, a private organization or a government, having different objectives to achieve. However, if the project has to thrive, it will need the sponsor's active support and involvement from the initiation stage, and to whom the project management should be accountable. The sponsor should have sufficient time to oversee the project. Ultimately, it is the sponsor who will carry the project forward from the beginning to a successful end.

2.3.4   Learning from Experience

The idea of learning from experience of earlier projects has always been a normal practice in the past. Galileo is believed to have employed case studies of failures for advancing and developing some of his path-breaking theories. This practice was continued for centuries when professionals and scientists almost always used precedents – both successes and failures – to buttress their ideas and decisions. Case studies from previous projects, from wherever these are available, have thus become recognized as a source of knowledge for developing ideas for the future and a symbol of well-thought-out project management for attaining success.

Of late with the increasing technological progress in project appraisal and execution system, the need for recording case histories of failures has assumed greater importance. However, in these days of litigations and arbitrations, there are many hindrances for candid discussions of project failures. As a result, frank analysis of errors during appraisal and/or execution stages is generally not found in technical literature.

2.4   APPRAISAL

Project appraisal is a process of assessing in a structured way, the viability of a project at the initial 'idea stage'. It involves examining various options from different perspectives using appropriate analysis techniques. In effect, the process is used for judging whether the project is acceptable or not to the stakeholders before pledging any resources to it. The stakeholders or the interested parties are investors, financiers, guarantors, licensors, etc.

2.4.1 PURPOSES OF PROJECT APPRAISAL

The main purposes of project appraisal are as follows:

- To define the project objectives and examine the options to attain these objectives;
- To ascertain the likely consequences of an investment by examining the available resources and their projected performance criteria. The appraisal indicates whether or not to invest in the project, considering different aspects;
- To identify the risk elements and probability of any unfavorable consequence.

2.4.2 APPRAISAL PROCESS

Appraisal process consists of five steps. These are as follows:

- Initial assessment;
- Problem definition and preparation of a long list;
- Reviewing the long list and preparation of a short list;
- Developing options;
- Comparing the options and selection of a project.

The process generally starts at the initial phase of a project so that the organization is in a position to make a decision of the amount of capital to be spent on the project or even discontinuing a project that is not viable or practicable.

2.4.3 EARLY APPRAISAL

In order to achieve success, it is necessary that once the alternative options have been developed in the 'idea stage', these options should be subjected to a detailed analysis from different perspectives for ascertaining their viabilities. These include:

- Market analysis;
- Technical analysis;
- Economic analysis;
- Financial evaluation;
- Environmental appraisal;
- Uncertainty and risk analysis.

These are discussed in the chapters that follow.

2.5   CONCLUDING REMARKS

All efforts should be made to achieve absolute success in any project, since failure is always costly and disappointing apart from being frustrating. In the preceding paragraphs, some of the important aspects of success in a project have been briefly discussed. Most of these have been drawn upon from experiences in previous projects. These need to be considered for going forward for success in projects. However, priorities may vary from case to case depending upon specific circumstances or problems.

BIBLIOGRAPHY

Baguley, P., 2008, *Project Management*, Chartered Management Institute/Hodder Education, London.
Batchelor, M., 2010, *Project Management Secrets*, Harper Collins, London.
Burk, R., 2007, *Introduction to Project Management*, Burke Publishing, UK.
Chandra, P., 1995, *Projects: Planning, Analysis, Selection, Implementation, and Review*, Tata McGraw-Hill Publishing Company Ltd., New Delhi.
Roberts, P., 2011, *Effective Project Management*, Kogan Page Ltd., London.
Smith, N.J., 1996, *Engineering Project Management*, Blackwell Science Ltd., London.
Westland, J., 2007, *The Project Management Life Cycle*, Kogan Page Ltd., London.

3 Market Analysis

3.1 INTRODUCTION

Market is a group of potential customers who are likely to buy products or services offered by an organization. Characteristic of a market is the act of exchange. Some people (consumers) acquire the products they need most, and other people (suppliers), who are capable of satisfying the need, provide these to the former. Thus, both sets of people feel satisfied by this exchange. Typically, both these sets of people like to maximize their own gains.

Market analysis provides ideas about the likely market demand of such products or services, their competitiveness, locations, etc. It encompasses the current as well as future scenario. Its objective is to analyze the potential market size and its value. It also provides an insight about how the products or services compare with those offered by the competitors in quality, price and availability.

3.2 MARKET ENVIRONMENT

The fast-changing technological development during the later part of the 20th century has changed the market trends throughout the world beyond recognition. Consequently, all organizations have to continuously review and monitor the scenarios that are prevailing outside as well as inside the organization to ensure that it adapts to the changing market scenarios. These scenarios are collectively known as market environment in business parlance.

Market environment consists of two categories. The first is external category, which is outside the direct control of the organization and is generally termed 'macro-environment'. The second category is termed 'microenvironment' and relates more to the specific segment of the industry, and the concerned organization has a certain degree of control over it. Although these two categories are generally specific and independent, they do have occasionally some influence over each other.

3.2.1 MACRO-ENVIRONMENT

Macro-environment of a market comprises the certain distinct and identifiable external factors that influence the prevailing market trends. These are often collectively known by the acronym PESTLE, which stands for the following factors:

- Political;
- Economic;
- Social;
- Technological;
- Legal;
- Environmental.

The salient features of these factors are briefly discussed in the following paragraphs.

3.2.1.1 Political Factors

Governmental plans, policies, legislation, etc. generally affect the market (customers) as well as the organizations. Consequently, the organizations need to consider the political risks arising out of national as well as international political issues. Otherwise, the business of the organization may be affected in both the short and long terms. Political instability may create economic instability and social unrest, leading to a reduction in production and in export of raw materials, as well as a decrease in the population's purchasing capacity. Additional concern is that political risks vary from time to time.

3.2.1.2 Economic Factors

Economic factors such as unemployment, taxation and interest rates. affect individuals and organizations alike. Given the current globalized economic scenario, the economic factor affects the working of the organizations not only within the boundaries of the country or region, but elsewhere also. Thus, companies with operations beyond their own boundaries should be aware of the risks involved in operations in the environments of different countries. These risks include terrorism, civil unrest, civil revolution and strikes. Also, under such circumstances, the local government may even confiscate the assets of a foreign organization without paying adequate compensation. Disproportionate income distribution and poverty are other economic factors that need serious consideration. A large proportion of the world's population is below poverty line. It does not have enough food every day and also does not have the opportunity to receive even primary education.

3.2.1.3 Social Factors

Macro-environment presents challenges to organizations in the social and cultural fronts also. Social factors such as income, occupation and sex affect the customer's demands. Culturally, language, values, beliefs, etc. can also influence these demands.

3.2.1.4 Technological Factors

Technological innovations can open up exciting new markets and opportunities. Currently, technology is perhaps the most dramatic force in our society, which is shaping our destiny. Computer systems and internet are increasingly being used in universities, public and private sector organizations. As a result, when computer or internet systems temporarily fail to operate (may be due to viruses), there is usually serious personal as well as financial impact. A chain of organizations and individuals may be affected by such failures. On the other hand, several organizations in several countries are carrying out researches on new medicines, new industrial materials, robotic technologies, etc., which may revolutionize several areas of product development.

3.2.1.5 Legal Factors

A combination of political and judicial influences often leads to legal actions, such as imposition of sanctions and embargoes or controlling of sale. For example, many governments around the world have introduced regulations for controlling the sale and advertisement of tobacco products.

3.2.1.6 Environmental Factors

Concern about air and water pollution has grown steadily over the last few decades in almost every country around the world. In fact, there is a threat of shortage of drinking water in many countries in the foreseeable future. Also, the potential threat of global warming coupled with severe weather patterns is likely to have a damaging effect on the community, their homes and livelihood. These environmental hazards are likely to affect the market environment.

3.2.2 Micro-Environment

Micro-environment consists of internal factors that focus on local issues, and can influence the organization's ability to meet the customer's needs. The main factors under this category are as follows:

- Employees;
- Suppliers;
- Local community;
- Customers;
- Intermediaries;
- Competitors.

3.2.2.1 Employees

Employees are the backbone of any organization and fountainhead of new ideas and knowledge. It is the people within it who drive it forward. Management and shareholders need to value the worth of this workforce in order to maintain a competitive position in today's dynamic market environment.

3.2.2.2 Suppliers

Dependable suppliers are vital for the successful operation of any organization. They furnish the resources required by the organization to produce its goods and/or services. It is therefore important to have a good working relationship between the organization and its suppliers. In this context, it should be borne in mind that the competitors may also be served by the same suppliers. Such suppliers should be dealt with diplomatically.

3.2.2.3 Local Community

This group includes neighborhood residents and community organizations who have some vested interest in the activities of the organization. It could be the residents of the locality, the media (newspapers, magazines, television and internet) or even the employees.

3.2.2.4 Customers

Customers are the most important actors in the organization's micro-environment. Without them, there would be no business. Thus, every organization must know its customers, where they come from, and what are their preferences, habits and attitudes. As they change, the organization must also change its strategy to continue to be in line with their preferences, habits and attitudes. Otherwise, the organization is bound to lose the customer.

Broadly, customer market may be categorized under five heads:

1. *Consumer market:* individuals and households who buy products and/or services for personal use.
2. *Business customer market:* organizations who buy goods/services as inputs for their own production for further processing and selling.
3. *Reseller market*: individuals and organizations who buy goods/services for selling these at a profit.
4. *Government markets* buy goods and services to produce public services or to transfer the same to other agencies.
5. *International markets* include all the above buyers situated in other countries.

3.2.2.5   Intermediaries

Intermediaries are the individuals or organizations that operate between the component- or service-suppliers, the manufacturers and retailers. There may be several intermediaries in a supply chain, and they play an important role in the chain. It is important for the organization to maintain good relationship with the intermediaries in order to ensure the smooth flow of components or services; otherwise, interruptions in the flow might affect the production process.

Intermediaries could be product distribution firms, marketing services agencies or financial organizations. Product distribution firms help the product to move through the channel so that it becomes available to the appropriate customer. These include wholesalers, retailers, agents and franchisers. Marketing research, advertising, etc. are done by the marketing services agencies. Financial intermediaries help the organization in financial transactions. These include banks, credit companies and insurance companies.

3.2.2.6   Competitors

In order to sustain in any industry, all organizations need to be competitive. They have to score a strategic advantage by positioning their products ahead of those of their competitors in the minds of their customers. If they do not, the customers will go to their competitors. However, the wining strategies for large organizations and small units may not often be similar.

3.3   MARKET ANALYSIS PROCESS

Market analysis process is a systematic method for collection of data, their in-depth study and interpretation for deciding on marketing strategies. The intensity and frequency of the study vary as per the requirement of individual projects. The exercise provides varied information to the stakeholders and assists them in appropriately evaluating the market behavior. The process comprises in-depth study primarily of the following aspects:

- Demand pattern, i.e., the potential size and composition of the market;
- Current and future demand;

- Product specification and quality as compared with those of the competitors;
- Degree of competition;
- Marketing strategy of the organization;
- Tariff level.

The relevant data for the foregoing aspects can be gathered in two ways, viz., through a market survey for the specific project (for primary data) or through existing reports and publications collected for other projects (for secondary data). The former, i.e., survey for primary data collection, is generally reliable since the data are collected to meet a specific objective.

3.3.1 Market Survey for Primary Data

Market survey provides information which may be used for the examination of different aspects of market. These include the following:

- Current demand and future prospect;
- Demands in respect of different segments of the market;
- Price variations and market sustainability;
- Purchase policy;
- Customer perspective;
- Product varieties;
- Distributive trade practices.

Although market survey has many advantages, its results can be impaired by a number of shortcomings in the system. These include the following:

- The samples may not be actual representative of the market;
- The questionnaire may not be adequate;
- The respondents may not be able to understand clearly the questionnaire;
- Consciously wrong answers by the respondents;
- Distortion of responses by survey staff;
- Inadequate analysis and interpretation of the data.

Market survey may be done through a census survey or a sample survey.

3.3.1.1 Census Survey

In census survey, *all* users of a particular product or service are covered (as, for example, all industries using a particular type of machine or all readers of a particular magazine). This survey, though very useful, often becomes prohibitively expensive. As a result, it is used primarily by large and important establishments like steel plants and defense establishments.

3.3.1.2 Sample Survey

In sample survey, only a sample (portion) of the users are contacted, and relevant information is gathered and examined. The result of the study is then extrapolated to a target population. A sample survey generally covers the following stages.

3.3.1.2.1   Define the Target Community

The target community should be defined clearly without any ambiguity. The target community should also be categorized into segments based on typical characteristics, such as income brackets and professions.

3.3.1.2.2   Select the Sampling Scheme

Selection of the sampling scheme is to be done from a few alternatives such as random sampling, cluster sampling, sequential sampling and systematic sampling. The reliability of the survey is directly related to the sample size; the larger the sample size, the more dependable would be the survey result.

3.3.1.2.3   Develop the Questionnaire

The questionnaire constitutes the primary instrument for bringing out the required information about the product or service. It should aim at an in-depth understanding of the product or service as well as of the various technical and specialized issues. Preparation of the questionnaire needs a thorough understanding of the product or service and its usage as well as attitude and motivation of the community. The questionnaire should be clear as well as thorough. It is general practice to try out the questionnaire in a pilot survey and modify the same in light of the problems faced.

3.3.1.2.4   Recruit and Train the Survey Personnel

Recruiting and right kind of training of survey personnel are time-consuming activities. It is imperative that competent survey workers with appropriate technical culture are recruited for this purpose.

3.3.1.2.5   Obtain Response of the Questionnaire

The respondents are to be interviewed either personally or by telephone or through internet. In developing countries, the first option is mostly adopted.

3.3.1.2.6   Study the Responses

The responses should be thoroughly studied, and irrelevant and/or unwanted responses should be eliminated. The remaining responses should be shortlisted for in-depth analysis.

3.3.1.2.7   Analyze and Interpret the Responses

The information gathered by the survey should then be analyzed and interpreted carefully. Suitable statistical method may be employed for this purpose, if deemed necessary.

3.3.2   EXISTING REPORTS AND PUBLICATIONS (SECONDARY DATA)

Information gathered from already available reports and publications of earlier projects is generally utilized as secondary database and plays a very important role during the initial stages of market analysis of projects. This exercise is generally started by utilizing the internet where a vast range of information is available online and often free of charge. The sources include reports of both private and public sector establishments, chambers of commerce, trade directories, websites of different government departments and industries. The secondary data available from existing reports and publications provide information about two aspects, viz., *market information* and *product features*.

3.3.2.1 Market Information

The existing reports and publications will primarily provide the following data:

1. Market size and structure;
2. Customers and suppliers;
3. Main products;
4. Market condition: new, matured or saturated?
5. Market response: satisfactory?
6. Methods of sales promotion: media, internet, e-mail, online, etc.?
7. Government policies, taxes, restrictions, etc.;
8. Legal aspects, viz., trade mark, copyright, patent, etc.;
9. Scope for expansion and diversification.

3.3.2.2 Product Features

The data available in this category include the following:

1. *Potential customers*: identification, location, etc.
2. *Product development*: new products or improvement of existing products necessary?
3. *Brand image*: satisfactory?
4. *Competitors*: identification, location, threat factor, pricing, distribution channels, new product introduced, etc.

3.4 STUDY OF MARKET CHARACTERISTICS

Based on the data gathered from Section 3.3 (Market Analysis Process), the following salient features of market characteristics can be arrived at:

- Market trend projection;
- Market segmentation;
- Price trend;
- Distribution and sales promotion;
- Consumer characteristics;
- Sources of supply;
- Socioeconomic policy of the government.

3.4.1 MARKET TREND PROJECTION

In this method, future demand of the product is assessed by extrapolating the trend of the past and existing demand into the future.

3.4.2 Market Segmentation

Market segmentation helps an organization to study the current as well as prospective market demands, since nature of demand varies from one segment to another. For consumer products and services, this activity is done to suit the requirements of different socioeconomic groups, age groups, occupations, etc. For industrial products and services, the customers are usually other companies or government departments.

The activity is helpful in market characterization since the nature of demand fluctuates from segment to segment. The segments may be grouped as follows.

3.4.2.1 Product Specification

For example, commercial vehicles may cover small- and medium-capacity vans, trucks/trailers or passenger-carrying buses of various capacities.

3.4.2.2 Geographical Area

Area-wise classification of consumers for the products will require regular after-sales services.

3.4.2.3 Consumers

They may be divided into two groups, viz., domestic consumers and industrial consumers. Domestic consumers may further be subdivided income-wise. Similarly, industrial consumers may be subdivided industry-wise.

It should be emphasized here that different strategies for market development will be required for different segments.

3.4.3 Price Trend

Reliable data relating to ex-factory price, insurance and freight expenses, average wholesale and retail price, etc. need to be made available for assessing the price trend of products or services.

3.4.4 Distribution and Sales Promotion

Distribution system varies from product to product. Raw materials, intermediate products and final products have different distribution channels. Sales promotional methods utilizing internet/media and discount/gift schemes also vary according to the nature of the commodity.

3.4.5 Consumer Characteristics

Consumers can be broadly characterized into the following categories:

- Residential location, age, sex, profession, income bracket, social background, etc.;
- Attitudes, preferences, habits, etc.

3.4.6   SOURCES OF SUPPLY

Information about existing sources of supply (foreign or domestic) is an important aspect of market characteristics. Possibility of replacement by another product needs to be studied as this might affect the price, quality, promotional efforts, etc. of the product.

3.4.7   GOVERNMENT POLICY

Socioeconomic policy of the government plays a vital role in influencing the demand and market of a product. This policy is indicated by:

- National plans;
- Import/export policies;
- Import duties;
- Export incentives;
- Subsidies;
- Credit controls;
- Industrial licensing;
- Tax on goods and services etc.

3.5   ASSESSMENT OF FUTURE DEMAND

Assessment of future demand is the next step after collecting information about the various market characteristics discussed in the previous section. There are several methods of assessment of future demands (forecasting). The three commonly used methods are as follows:

- Qualitative methods;
- Time-series projection methods;
- Casual methods.

In the present text, only brief description of these methods will be given. For further information, a study of specialist literature is recommended.

3.5.1   QUALITATIVE METHODS

In these methods, evaluation by experts is utilized to translate qualitative information into quantitative estimates. Under this category, two methods are important.

3.5.1.1   Jury of Executives Opinion Method

This method is a popular one. In this method, the opinions of a group of executives on expected future sales are pooled to assess the expected future sales and their values.

3.5.1.2   Delphi Method

In this method, the information is gathered not by face-to-face interaction, but through mail survey via a set of questionnaire and the responses thereof.

3.5.2 Time-Series Projection Methods

Analysis of historical time series is used to generate demand forecasts. Under this category, three methods are important.

3.5.2.1 Trend Projection Method

In this method, trend of consumption in the past is extrapolated to assess the future consumption.

3.5.2.2 Exponential Smoothing Method

In this method, the observed errors are taken into account in assessing future consumption.

3.5.2.3 Moving Average Method

In this method, simple or weighted arithmetic average of previous consumption is considered for computing future consumptions.

3.5.3 Causal Methods

In causal methods, forecasts are developed on the basis of cause–effect relationships. Under this category, a more analytical approach based on a detailed quantitative manner is adopted to develop the forecasts. The important methods under this category are as follows:

- Chain ratio method;
- Consumption-level method;
- End-use method;
- Economic method.

3.5.3.1 Chain Ratio Method

This is a simple analytical method in which the total estimated market demand is multiplied by a factor to arrive at the sales potential of the organization.

3.5.3.2 Consumption-Level Method

The method is useful when the product is directly consumed and measures consumption level on the basis of income elasticity of demand and price elasticity of demand.

3.5.3.3 End-Use Method

The method is suitable for forecasting demands of intermediate products and involves the following steps:

Step 1: Identify the possible use of the product
Step 2: Estimate the consumption coefficient of the product for various uses
Step 3: Project the output levels of the consuming industries
Step 4: Assess the demand of the product

However, since the consumption coefficient may vary from time to time due to recent trend of fast technological changes and improvements, this method should be used judiciously.

3.5.3.4   Econometric Method

This is a sophisticated forecasting tool. In this method, the forecasting involves estimating quantitative economic relationship derived from economic theory. This is an expensive and data-demanding process, which is a constraint for using this method for day-to-day forecasting purpose.

3.5.4   UNCERTAINTIES IN ASSESSMENT OF DEMAND

Uncertainties in the assessment of demand arise from the following primary sources:

- Inadequate quality and quantity of data about past and present market condition;
- Limitations to handle unquantifiable factors;
- Unrealistic assumptions in the forecasting methods;
- Requirement of excessive data for more advanced methods of demand assessments;
- Technological change and development in the recent past, e.g., new product which is comparatively more efficient and economical than on existing product may penetrate into the market of the existing product;
- Change in government policies, e.g., providing licenses/incentives to new companies, restricting/liberalizing of import/export of certain commodities, etc. have significant effects on the business situation;
- Change in international political and/or socioeconomic situation may also have considerable effect on the industrial scenario;
- Discovery of alternative sources of raw materials;
- Climatic unpredictability and natural disasters, e.g., earthquake, Tsunami, flood, drought, directly or indirectly affect the demand of a wide range of products.

3.6   MARKET STRATEGY AND PLANNING

Market strategy and planning may need to be developed in detail for penetrating the market to the desired extent. The concerned activities include the following:

- Price setting;
- Distribution;
- Promotional activities;
- Customer service.

3.6.1   PRICE SETTING

While setting the price of a product or service, several balancing factors need to be considered. These include the following.

3.6.1.1   Customers

The organization needs to pose a question as to how much the customer is willing to pay for the benefit he or she gets out of the product or service offered. This perspective may, however, change in the future. Therefore, to achieve proper assessment on the issue, appointing a permanent research team is necessary. This may be impracticable for many organizations for lack of resources as well as time.

3.6.1.2   Competitors

Quality and price level of the product or service offered by the competitors quite often influence the price setting strategy of an organization. Unless these are competitive, customers will move away to competitors' product or service. Thus, it is essential that any organization must ensure that its products or services are more attractive than those of its competitors.

3.6.1.3   Costs

Every product or service is produced at a cost and the organization must cover its cost, including overhead expenses *plus* profit margin while setting the price.

3.6.1.4   Corporate Objectives

Corporate objective is an integral part of a firm's price setting policy. For example, a firm with a strong fund position may like to set the price low (and make smaller profit initially) in order to capture the market and outprice the competitors. Alternatively, if the prices are set high, the customers may not be willing to purchase the item, forcing the firm to lower the price. But if proper margin is not acceptable, it may be difficult for the firm to lower the price. The process of price setting in such cases will be a difficult balancing act.

3.6.2   DISTRIBUTION

Distribution methods are important for reaching the firm's product into the hands of the customers. There are several methods normally used for this purpose. These include the following:

- Field sales where the sales persons personally visit the prospective client's premises to deliver the product for sale;
- In house sales department which uses direct methods like internet and telephone for making the sale;
- Employment of intermediaries or distributors for distributing the product to the retailers who in their turn make the sale to the ultimate user;
- Self-service retail method;
- Full-service retail distribution channel.

3.6.3   PROMOTIONAL ACTIVITIES

Promotional activities broadly include advertising and branding. In recent years, digital advertising has taken a bold step in outstripping other types of promotional activities. Market research may help in the planning and implementation of such exercises.

3.6.4 CUSTOMER SERVICE

Customer service has a strong influence on long-term market success of a product as well as on the reliability of the producer. The service starts from the time of installation of the product and continues thereafter also, in educating the customer, in providing warranties and satisfactory after-sales service etc. Availability of sufficient service outlets for this purpose is very important.

3.7 CONCLUDING REMARKS

Market analysis provides timely marketing information so that the business risks are reduced, sales opportunities are increased, current and future problems are identified, and appropriate remedial actions are taken. Also, this exercise suggests the likely selling price of the products or services over a specific period of time.

An effective market analysis procedure should follow a logical and systematic approach, comprising several distinctive activities. These activities have been briefly discussed in the foregoing sections.

BIBLIOGRAPHY

Bhattacharjee, S.K., 2008, *Fundamentals of PERT/CPM and Project Management*, Khanna Publishers, Delhi.

Chandra, P., 1995, *Projects: Planning, Analysis, Selection, Implementation, and Review*, Tata McGraw-Hill Publishing Co. Ltd., New Delhi.

Gould, R., 2012, *Creating the Strategy: Winning and Keeping Customers in B2B Markets*, Kogan Page Ltd., New York.

Hague, P., Hague, N. and Morgan, C.-A., 2013, *Market Research in Practice*, Kogan Page Ltd., London.

Market Analysis and Marketing, 2005, United Nations Industrial Development Organization, Vienna.

Oldcorn, R. and Parker, D., 1996, *The Strategic Investment Decision: Evaluating Opportunities in Dynamic Markets*, Pitman Publishing, London.

4 Technical Analysis

4.1 INTRODUCTION

Technical analysis provides a broad view of the technical feasibility of a particular project idea in order to establish that the same is rational, well founded, appropriately engineered and follows accepted standards. This study should be taken up early and during the planning stage itself of a project. Different alternatives involve different issues that are to be considered in the study. These issues generally vary from project to project.

4.2 OBJECTIVES OF THE STUDY

At the preliminary stage, there are usually a number of alternative technical solutions from which the most appropriate one in terms of viability, operating features and costs needs to be selected. This is the primary objective of technical analysis. Furthermore, for capital-intensive technologies, there are often capacity limits, and the one with most favorable operational characteristics should be selected.

The other objective of this study is to ensure that the entire system of organization and technology operates in unison to achieve the purpose of the project. Therefore, technical analysis should also attempt to identify and correct weaknesses that could adversely affect the project.

Last, but not the least, the objectives and priorities of the sponsor(s) must be considered along with technical aspects as these would form the basic frame of reference for ultimate decisions. Thus, in a private sector, profit would be the primary motive force (except in certain special instances), while in public sector units and public utilities, social benefit is more important than profit making. In either case, the project analyst has to focus on the objectives of the organization while carrying out his activities.

4.3 MATERIAL INPUTS

It is necessary to identify, quantify and evaluate inputs like:

- Raw materials;
- Processed industrial materials and components;
- Other supplies.

4.3.1 RAW MATERIALS

Raw materials generally include the following items that need to be studied properly.

4.3.1.1 Farm Products

These are agricultural products. Points to be studied are current availability and future potential of the product, current area of land available and future prospect of increase of land area, yield per acre, etc.

4.3.1.2 Mineral Ores

The aspects to be investigated are the physical, chemical and other properties of the ore available and whether the ore has (or likely to have in the future) a demand. The study should also provide an idea about the location, size and depth of deposits as well as viability of underground or opencast mining.

4.3.1.3 Other Products

Data on other products such as animal/livestock, forest products and marine products are not generally readily available. As such, specific survey may be warranted to get appropriate data on these products.

4.3.2 PROCESSED/SEMI-PROCESSED COMPONENTS

Processed and semi-processed components, subassemblies, etc. are also important inputs for any industrial venture. It must be ensured that these inputs are of right quality and of competitive price throughout the duration of the project life cycle. In case these are not easily available, it may be necessary to explore new sources of supply or develop sustainable sources of supply for long-term advantage. If necessary, the possibility of import may have to be considered.

4.3.3 OTHER SUPPLIES

Apart from the items mentioned above, other items like chemicals, additives, packaging materials, oil and grease are required in any industrial project. These also need to be considered in the technical feasibility study.

4.4 TECHNOLOGY/MANUFACTURING PROCESS

It is common experience that a particular product or service may be acquired by using two or more alternative technologies. For example, electricity-generating units may be thermal plants, hydroelectric power plants, nuclear power plants, solar energy plants, wind energy plants, etc., using different sources of energy. Also, the plants could be fully automatic, or semi-automatic. These aspects come within the preview of technology or manufacturing process.

The technology to be chosen for a project should be most up-to-date incorporating the latest developments in order to avoid or at least minimize the chance of becoming technologically out-of-date in the near future. It should also be flexible to adjust with or absorb newer technologies which may be available in the future.

Choosing of the technology from among a number of alternatives for carrying out a production program should be made after careful study of a variety of factors. These factors are briefly discussed in the following paragraphs.

4.4.1 PRINCIPAL INPUTS

The quality of available materials and other required resources should satisfy the required specifications. The cost of these materials and resources would be one of the major deciding factors.

4.4.2 PRODUCT MIX AND QUALITY

These must be of acceptable standard.

4.4.3 PLANT CAPACITY

This should be compatible with the production technology.

4.4.4 SCARCE RESOURCE

In case scare resources are required as inputs for the technology, directives for regulatory authorities may preclude the use of such resources for the manufacturing process, thereby excluding these technologies from probable choice.

4.4.5 LABOR VERSUS CAPITAL COST

The cost of labor as compared with the cost of capital consumed per unit output is a factor to be considered for technology choice. Labor-intensive technologies normally have advantage of production flexibility, while capital-intensive ones have a better quality control system.

4.4.6 RELIABILITY

A reliable technology has the advantage of minimal downtime for repair, thereby minimizing the cost of production.

4.4.7 COST OF TECHNOLOGY

Technology, whether developed internally or acquired from outside source, costs money. Effect of such investment and production cost over a time period should be examined carefully.

4.4.8 ABSORPTION OF TECHNOLOGY

In case the existing project staffs are not conversant with the technology, expatriate staff may have to be recruited. However, this alternative is costly and may also be considered while considering this option.

4.4.9 ENVIRONMENTAL IMPACT

Impact of the technology on the environment is a major determinant of technology choice. Measures necessary to mitigate the impact to acceptable level of existing regulations and also the possibility of future regulations need to be considered in detail.

4.4.10 SUSTAINABILITY UNDER LOCAL CLIMATE CONDITIONS

This aspect is important. For example, steel becomes brittle at low temperatures. Therefore for areas where temperature may be very low during winter, technologies using steel should be avoided.

4.4.11 SERVICES

For advanced technologies, intensive and comparatively complex services are required. Care should be taken to ensure that interruptions to such services are avoided.

4.4.12 LOCAL REGULATIONS

Often local authorities offer incentives for projects that are likely to upgrade the technological level of the area. Dialogue with such authorities should be initiated on a priority basis.

4.5 PRODUCT MIX

The choice of product mix is essentially guided by market requirements for satisfying diverse series of customers, with different tastes, needs and bearability of price range. Flexibility in respect of product mix enables the producer to stay and grow in the changing market conditions. For example, a biscuit- or snack-manufacturing organization may market two types of packages: a small package for low-cost one-time users and a larger package with price reduction offer for family use. Quality aspect must enjoy paramount interest while considering this aspect to ensure higher profitability. Careful analysis of further investment required for this purpose is another aspect to be considered prior to taking final decision.

4.6 PLANT CAPACITY

Plant capacity or production capacity normally refers to the volume or number of units that can be produced by a manufacturing plant during a particular time frame. This capacity is attainable under normal working conditions and depends on the following factors.

4.6.1 TECHNOLOGICAL CONSIDERATIONS

A certain type of plant of a minimum economic size will be required to achieve a particular production capacity (such as so many units per day). In case the available plant cannot meet the requirement, an alternative technological solution will have to be resorted to.

4.6.2   INPUT CONSTRAINTS

In many countries, particularly in developing countries, constraints such as limited power supply, scarce raw materials and foreign exchange restrictions are quite frequent. Some constraints may be seasonal. These need to be borne in mind while considering capacity of a plant.

4.6.3   ENVIRONMENTAL CONSIDERATIONS

Temperature, atmospheric pressure and humidity can have adverse effect on the running of machineries and equipment. At high elevation, average atmospheric pressure is reduced; this can affect the performance of turbines and combustion processes. Also, steel tends to get brittle at very low temperature; this may affect the machines. Performance of heat-related processes such as condensers and evaporators is dependent on ambient temperatures.

4.6.4   RELATIONSHIP BETWEEN MANAGEMENT AND WORKERS

This is an important factor that contributes to the efficiency of the plant capacity. Cordial relationship between the management and workers brings out the best from both the parties in question and contributes to the achievement of the feasible normal capacity.

4.6.5   MAINTENANCE

The amount of downtime due to maintenance of a plant for trouble-free operation is another factor for achieving the feasible normal capacity.

4.6.6   MARKET CONDITIONS

Anticipated market for goods or services has an important influence on the decision of the capacity of the plant. If the market is likely to increase, a plant of higher capacity would be preferable. However, if the growth is uncertain, it is advisable to commence the project with a lower capacity and additions to the capacity can be worked out in case the demand tends to increase.

4.6.7   RESOURCES OF THE ORGANIZATION

A plant cannot opt for a scale of operation beyond its operational or monetary capacity. Thus, resources of the organization, in both operational and financial areas, will influence the capacity of the plant.

4.6.8   GOVERNMENT POLICY

Regulatory policy of the government influences the plant capacity of a unit. This policy may vary from country to country and also time to time. Feasible normal capacity is subject to such state policies.

4.7   CHOICE OF LOCATION

Location signifies a broad geographical area such as urban/rural area, industrial zone and coastal area where a project is being considered to be set up. Various aspects are to be taken into consideration for selecting a particular location.

4.7.1   PROXIMITY TO RAW MATERIALS

It is important that location of any project should be close to the source area of the raw materials. Apart from other practical advantages, this would substantially reduce the costs for transporting the raw materials from their source area to the plant. As for example, a hydroelectric power station needs a perennial source of water for its sustenance. Also, a steel plant needs supply of iron ores at a reasonable distance away from the plant site. A cement plant, likewise, needs supply of limestone from nearby source. In all such cases, location of the plant near the source of raw materials would be a great advantage.

4.7.2   PROXIMITY TO MARKET

As in the case of raw materials, it would be an advantage to locate the plant close to the product markets as well.

4.7.3   AVAILABILITY OF INFRASTRUCTURAL FACILITIES

Before the location of a plant is decided, it is necessary to make sure that certain infrastructural facilities such as transport, power, water and modern communication system are available near the proposed location.

4.7.3.1   Transport

Nature and condition of existing roads, railway facilities (like station and yard), road and rail bridges, air transportation facilities, inland water and sea linkages, including port facilities, need to be assessed before the location is finalized.

4.7.3.2   Power

Availability of uninterrupted power supply with minimum voltage fluctuation and favorable tariff are the important aspects of infrastructure availability. Also important is the level of investment required for connecting the plant with the network of power-supplying agency.

4.7.3.3   Water

Water requirement for a plant varies according to the plant capacity and technology adopted. This can be satisfied either by drawing from external sources like public utilities or internally from surface/subsurface sources, depending on the quality, dependability and costs.

4.7.4   GOVERNMENT POLICIES

Government policies often dictate the choice of locations, both for public sector and for private sector projects. Thus, due to some broader policies for the dispersion of particular industries, due to, say, congestion, the government may direct such industries to be located away from urban areas. Similarly, the government may offer incentives for establishing industries in backward areas, by way of subsidies, concessional financing, tax reliefs and similar benefits.

4.7.5 AVAILABILITY OF SUPPORTING INDUSTRIES

One important consideration for choice of location is availability of supporting industries. It is a common practice for a plant to get some production operations done from outside sources by subcontracting. For this purpose, it is essential to study whether suitable industries are operating in the surrounding areas to supplement the resources of the plant.

4.7.6 ENVIRONMENTAL POLLUTION

A project may emit unwelcome gases, and produce detrimental solid and liquid wastes; it may cause noise, heat and vibration. These all cause environmental pollution. This aspect should be considered, and costs for reducing such environmental pollution to acceptable levels or if necessary shifting the project to alternative locations may be considered in the study.

4.7.7 HUMAN RESOURCES

The other aspect that gives edge to a particular location is the availability of skilled workforce at competitive rates. Thus, labor-intensive projects like textile and jute mills thrive in areas where cheap labor is available. Similarly, industrial units normally grow where skilled, semi-skilled and unskilled labors are concentrated. Depending on the level of technology used for a particular unit (automatic, semi-automatic, etc.), the technical appraisal is done accordingly. The other factor that needs to be considered is the prevailing state of industrial relationship between the management and the labor in terms of frequency and intensity of strikes, lockouts, etc.

4.7.8 CLIMATE

Areas with history of heavy rainfall, flooding, heavy snows and severe cold or heat, wind, etc. have adverse effects in the selection of any project. For such areas, additional funds for dehumidification, heating, air conditioning, etc. would probably be needed. Special attention must be given if an area comes within heavy earthquake zone.

Pleasant climatic conditions with natural and scenic beauty may attract tourists, and thus, such a location may boost the growth of health resorts and tourist complexes.

4.7.9 LIVING CONDITIONS

In addition to the forgoing considerations, a project should have adequate communication facilities like internet connection. Suitable arrangements for disposal of solid waste, liquid effluent, etc. are also to be investigated. Social amenities like cinema halls, multiplexes, theatre halls, eateries, parks and playgrounds also form part of infrastructure development. Other considerations also include cost of living, availability of accommodation for workmen, proximity to market, healthcare facilities and availability of educational institutions. A rational analysis of such requirements and their implications in terms of time, money and resources is imperative at the appraisal stage itself, to avoid surprises at a later date.

4.8   SITE SELECTION

While location signifies a broad geographical area, site refers to a particular area of land where a project can be set up. Generally, once the location is selected, a few alternative sites are considered and scrutinized in respect of size and shape of the area etc., as well as cost of land and civil work for the development of site.

4.8.1   Considerations for Site Selection

Selection of a particular site from several alternatives calls for a systematic analysis. The analysis should be carried in the following sequence.

4.8.1.1   Size and Shape of the Site

It must be ensured that layout of buildings, structures, plants and equipment storage areas, parking facilities, etc. are satisfactorily accommodated within the size and shape of the available site.

4.8.1.2   Environmental Sensitivity

A site may be situated in an environmentally sensitive area such as wetland or in a natural habitat for endangered species where industrial development is prohibited. Such a site has to be rejected from reckoning forthwith.

4.8.1.3   Future Expansion

There should be scope for additional space for future development, viz., plant extensions, and other add-on facilities.

4.8.1.4   Soil Condition

Soil condition, particularly bearing capacity for foundations of new structures, should be examined before finally selecting a site.

4.8.1.5   Layout Plan

A detailed layout plan should be drawn showing arrangements of all the buildings, structures and positions of plants and equipment with important dimensions and levels. This should be considered as a record for a particular alternative site.

4.8.2   Estimated Investment Cost

On the basis of the layout for each particular alternative site, estimates for investment cost are to be computed on the basis of the following heads.

4.8.2.1   Cost of Land

Cost of land may vary from site to site, even within the same broad location. Sites close to urban areas normally cost more than those in the rural areas. At the same time, cost of land in specified areas specially developed by the government for industrial purposes may be available at concessional rates.

4.8.2.2   Civil Works and Other Expenses

Costs involved in construction/relocation of existing structures relate primarily to:

* Excavation, concrete works, masonry, roofing, steel/aluminum sheet works, etc.;
* Special civil engineering activities like soil investigation, soil consolidation, pile foundations, drainage, ramps, chimneys, silos and foundations for heavy equipment;
* Specialist items like carpentry, joinery, steel works, plastering, glazing, tiling, flooring, asphalting and painting;
* Technical installations like heating, ventilation, air conditioning and plumbing;
* Utility supplies and distributions, e.g., water, electricity, communications, steam and gas;
* Traffic installations like yards, roadways, parking areas, railway tracks and sheds for bicycles;
* *Landscaping*: trees, plants, grass, etc.;
* Others like security installations.

4.9   SELECTION OF MACHINERY AND EQUIPMENT

One of the major engineering responsibilities in a project is the selection of machinery and equipment. These should be selected primarily to meet the requirement of the production and should be commensurate with the surrounding operating conditions.

4.9.1   Technology

It is common practice that in case a technology is acquired through a turnkey agreement, the machinery and equipment are selected by the technology provider. In other cases, the project designers select these to suit the technology. If in-house technical expertise is insufficient, consultants from external source (inside or outside the country) may have to be employed for providing the required technology.

4.9.2   Plant Type and Capacity

Requirement of machinery and equipment is influenced by the type and capacity of any particular plant. Consequently, machinery and equipment required for a process-oriented industry will differ from those required for a manufacturing industry.

The machinery and equipment required for a project may be categorized as follows:

- Process equipment;
- Mechanical equipment;
- Electrical equipment;
- Instruments;
- Control system;
- Spare parts, tools, etc. for running the plant and maintenance purpose.

4.9.3   OTHER ISSUES

While finalizing the selection of plant and machinery, various issues need to be considered, which include the following:

- Quality assurance plan (QAP);
- Proven and up-to-date technology;
- Reputation of suppliers;
- Delivery/payment schedule;
- Performance guarantee;
- Equipment life;
- After-sales service;
- Availability of sufficient power to run electricity-incentive plants;
- Transportation of heavy equipment;
- Initial difficulties in running technology-intensive equipment, e.g., computer numerically controlled (CNC) machinery;
- Import policy of the government in case the machines are to be imported from abroad.

4.10   CONSTRUCTION ACTIVITIES

Development of the site and construction work are associated with any industrial enterprise. The nature of the industry, size and location of equipment and machineries, the manufacturing process, etc. govern the layout and sizes of the structures and buildings. First, the site needs to be leveled and unnecessary structures/buildings removed. Relocation of existing structures like cables, pipelines, power lines, water lines, roads, railway sidings, communication network (like telephones and internet) is to be done. Construction of new work includes factory buildings, laboratories, control rooms, administrative buildings, stores, warehouses, healthcare centers and staff quarters.

Apart from the foregoing activities, construction works also include treatment of factory waste materials, effluents, supply and distribution of utilities like water, electric power, gas, roadways parking areas, railway tracks, garages and lighting facilities.

BIBLIOGRAPHY

Bhattacharjee, S.K., 2008, *Fundamentals of PERT/CPM and Project Management*, Khanna Publishers, New Delhi.

Chandra, P., 1996, *Project, Planning, Analysis, Selection, Implementation, and Review*, Tata McGraw-Hill Publishing Co. Ltd., New Delhi.

Investment Project Preparation and Appraisal: Teaching Materials, 2005, United Nations Industrial Development Organization, Vienna.

5 Economic Analysis

5.1 INTRODUCTION

Modern concept of an enterprise calls for a balance of interest of the participants such as the employees, project sponsors, customers, suppliers, the state and the public at large. This concept ensures that the business activities are beneficial to the entire society (economy). The governments, in both free and controlled economies, protect this concept by enacting appropriate regulatory laws to ensure that all the sectors of the society – public, private, small or big – work for the overall benefit of the society. It is therefore logical that the pluralistic nature of the interests can be achieved if the economic analysis is done from a national point of view, rather than restricted to a regional issue or to a particular wing of the administration. In other words, consequences to all persons or groups of persons in the country should be considered in the analysis.

The method is generally used prior to commencement of project activity, early enough to be able to focus on the desired objectives of the project. However, it is a practice to use it during the project duration and also after the project is completed in order to ascertain the degree of its success during implementation and on completion. The technique is normally applied for the analysis of public investments. However, this is used for private enterprises also, where national socioeconomic considerations nowadays play an important role in investment decisions within the framework of broad strategic options at the macro-level.

5.2 FUNDAMENTAL CONCEPTS OF ECONOMIC ANALYSIS

5.2.1 PROLOGUE

It is a common practice that the costs of development projects and their maintenance are borne by the state, whereas the benefits arising out of these are enjoyed by the general public. Consequently, it is logical that economic analysis of such projects should be carried out from a national point of view rather than a regional or sectorial government department. This implies that the analysis should be done keeping in view the interests of every citizen of the community.

Financial and economic appraisals are somewhat similar to each other, because both assess profitability of investment. However, the undergoing concept of financial profit is not the same as economic profit. While financial profitability indicates commercial and monetary viability of a project, economic profitability measures the real worth of a project from the point of view of the society.

In effect, financial appraisal is a technique for estimating the rate of return of the investment in monetary terms. Economic appraisal, on the other hand, estimates the return on investment on all the segments of the society as distinct from a project entity. The basic difference between financial and economic appraisal is that the former assesses profitability based on market prices, while the latter measures the net impact of the investment on the national economy.

5.2.2   DIFFERENCE BETWEEN FINANCIAL ANALYSIS AND ECONOMIC ANALYSIS

Financial analysis deals primarily with the means of financing a project (by levying toll or floating bonds etc.), and its monetary profitability. Financial analysis is thus concerned with the source of financing, availability and allocation of funds. For example, in a toll road bridge, the costs and expected toll collection will decide whether the project will be viable from financial considerations and management has to take a decision to construct it or not depending on whether the toll collection seems attractive enough. This analysis is financial in nature. Economic analysis, on the other hand, concerns with the consequences of such decisions to all sections of the community, and this will be necessary for establishing the economic viability of the project. In this connection, it needs to be clearly understood that even if the economic viability is established, the authorities may still decide not to take up the project due to financial constraints, such as lack of funds or unattractive returns.

5.2.3   STUDY OF THE FUTURE

Economic analysis is not concerned with the past events or investments. It is essentially a study of what would happen in the future. Consequently, economic analysis should estimate future results of any development project.

5.2.4   ALTERNATIVE SOLUTIONS TO BE CONSIDERED

It is mandatory that a number of possible alternatives are evaluated and the most attractive option is selected (including 'do nothing' option).

5.2.5   QUANTITATIVE AND QUALITATIVE APPROACHES

Some consequences can be quantified into monetary terms, and some cannot. In economic analysis, effects that can be quantified monetarily should be included in the analysis. Non-quantifiable consequences are to be specified and presented to the decision maker for his consideration. For example, the impact of a project on the safety aspect of the public is an extremely important consideration. Since this cannot be quantified in monetary terms before it happens, the concerned engineer should highlight in qualitative terms, the effects of such future safety aspect.

5.2.6   ANALYSIS PERIOD

For minor private projects, it is common to have a short analysis period (say 2–5 years). For major public sector projects, the analysis period may extend to 15–20 years or even more beyond the completion of such projects.

5.2.7 Common Time Datum

It is customary to have all future costs and benefits to a common time datum since these occur at different points of time. For this purpose, discounted cash flow (DCF) procedure is followed. This method is discussed in a separate chapter dealing with financial aspect of project.

5.3 PURPOSES OF ECONOMIC ANALYSIS

The main purposes of economic analysis of a project are as follows:

a. Ensuring efficient allocation of the resources towards the economy of the community at the national, regional and local level within the overall development plan. Consequently, the efficiency with which the resources are used can have an impact on the performance of the economy and prosperity of the country;
b. Ranking of different options considering scarce resources in order of priority;
c. Assisting in phasing the development program over a period of time considering availability of resources;
d. Comparing the different options and selecting the most attractive one;
e. Determining whether the option is worth the investment at all;
f. Evaluating alternative strategies in respect of specifications, design standards and other parameters.

5.4 DISTINCTIVE FEATURES OF ECONOMIC STRATEGIES FOR PUBLIC SECTOR AND PRIVATELY OWNED PROJECTS

Public sector projects are approved, named and operated by the government or its agencies (public utilities), unlike privately owned projects where investment decisions are taken primarily by the private stakeholders.

Some distinctive features of economic strategies in public sector and privately owned projects are discussed below:

a. In a public sector project, goods, services, jobs, etc. are provided at a 'no-profit-basis'. This is generally not the case for a privately owned project;
b. The source of capital in a public sector project is primarily taxation, while in private sector, it is generally private investors, or promoters. In case of public sector projects, occasionally financing is done by self-liquidating bonds issued to the public or by subsidies or loans from financial institutions;
c. It is common to have multipurpose projects in the public sector, such as flood control, irrigation and power generation. This is seldom the case in a private project;
d. Life span of private projects is generally shorter (say 5–10 years), while a public sector project may run for longer period (20–50 years);
e. Political pressure in public projects may be much more than that in a privately owned project.

In view of the foregoing, economic analysis of projects in the public sector may not always be possible to be done exactly in the same manner as is done in the case of privately owned projects. This creates problems for the public in general, as well as for the decision makers and managers in the public works.

5.5   SOCIAL AND FINANCIAL COSTS AND BENEFITS

Appraisal technique in private sector is primarily concerned with maximization of profit and consequently private wealth. Thus, only private costs and returns of investment which occur directly to the investor are taken into account. However, private monetary costs and benefits of a project do not necessarily reflect its true social costs and benefits. For example, while undertaking appraisal of an industrial project, the investment required to construct the plant, the costs of equipment, raw materials, labor, overheads, etc. and the revenues arising out of the operation are generally considered. However, broader social and environmental consequences of constructing such a plant are not considered. These include both beneficial and detrimental effects. Examples of beneficial effects are increased employment potential and effect on balance of trade position (if the product is exported). On the other hand, environmental pollution, effect on local property prices, etc. contribute to the detrimental effects. Social costs and benefits are quite often inclined to differ from financial costs and benefits. Some of the major sources of differences are discussed below.

5.5.1   IMPERFECTIONS IN MARKET PRICES

Market prices which form the basis of monetary costs and benefits are not always under perfect competition. Under the circumstances, they do not always reflect the actual social values. Some common examples of imperfections are as follows:

- *Rationing system*: the price paid by the consumer for consumables purchased in this system is often less than the competitive market rate, i.e., the items are subsidized.
- *Minimum wages rate system*: in this case also, the prescribed minimum wages paid to labor are more than those which would have been prescribed in a free labor market.

5.5.2   EXTERNALITIES

Projects may have beneficial external effects. These are considered as social benefits, while from monetary point of view, they may be ignored because they do not contribute anything to the sponsors. A new road bridge connecting two towns is considered a social benefit for both the towns, but may be ignored from the monetary benefit point of view to the sponsors, unless, of course, a toll system is introduced. Similarly, the same bridge may give rise to vehicle congestion, and thereby have a harmful external effect like environmental pollution. This effect is relevant from social point of view, but may not adversely affect the sponsors from monetary point of view. Needless to add such externalities are very relevant for economic analysis of projects, since consequences of such economic activities affect other parties without this being reflected in market prices.

5.5.3 TAXES AND SUBSIDIES

Though taxes and subsidies are costs and benefits in monetary terms from the private point of view, these are considered as transfer of payments from social point of view and are often ignored in economic analysis.

5.5.4 DISTRIBUTION OF BENEFITS

A private sector organization may not be concerned about how the benefits are distributed among the various groups in the society. From the point of view of the society, however, such distribution of the benefits to a particular section of the population may be considered more valuable than the same to other sections. For example, the state may decide to promote an adult education program or a balanced nutrition program for schoolchildren, even though these may not be considered necessary from private point of view. Nevertheless, these are important from the social perspective.

5.5.5 CONCERN FOR SAVINGS

In economic analysis, savings and investments are considered as more valuable than consumption. A private firm does not bother about differential valuation on savings and consumptions. From a social point of view, however, difference between consumption and savings is relevant since saving will lead to investment, particularly in capital-scarce countries.

5.5.6 SHADOW PRICING

Actual expenses and revenues from goods and services do not always reflect the measurement of the costs and benefits to the society. For evaluating such expenses and revenues in terms of social costs and benefits, adjustments are required in the expenses and revenues in order to make them reflect their proper market value. These adjustments are known as shadow pricing. The principle of shadow pricing applies to both the cost stream and the benefit stream. Shadow pricing technique is adopted basically for overcoming the following difficulties.

5.5.6.1 Imperfect Pricing Mechanism

In many countries, particularly in developing countries, because of lack of perfect competition, domestic pricing mechanism does not operate perfectly. Relative costs, benefits and scarcities are not always reflected correctly in the pricing mechanism. This is due to direct and/or indirect influences on the demand and supply of goods and services emanated from operations by government agencies. As a result, domestic prices are not in line with the rates at which they could be traded in the international market. Such difficulties can be overcome with the help of shadow pricing technique.

5.5.6.2   Variation in Wages Rates

Due to unemployment problems in many developing countries, the wages of labor, particularly in the non-organized sectors, are often regulated by the government and tend to be lower than the logical level. This is primarily due to mass underemployment and unemployment at existing wages rates. In the organized industrial sector, on the other hand, labor forces have strong trade unions and wages tend to be higher than the opportunity cost of labor. Thus, actual wages of the labor need to be adjusted for calculating the labor cost for the purpose of shadow pricing.

5.5.6.3   Disparity in Interest Rates

Cost of capital is generally indicated by the interest rate. In developing countries, the majority of the population are impoverished with low savings level and therefore do not have the propensity to save and invest. Moreover, the relationship between supply of capital and interest rates prevalent in the country is minimal. There is thus wide disparity between interest rates prevailing in different geographical areas. To overcome such problems, shadow rate of interest is estimated based on the interest rates paid by private investors.

5.5.6.4   Disparity in Exchange Rates

In general, the developing countries suffer from adverse balance of payment in the foreign exchange area. As a result, the rate of foreign exchange tends to be lower in the open market than the official rate. The problem is solved by fixing a higher exchange rate than by fixing the official exchange rate in the project. This is tantamount to attaching weight to the cost of foreign exchanges in the project.

5.5.6.5   Inflationary Forces

Some projects take considerable time from conception to completion. During this period, due to inflationary trend, cost of labor, material, equipment, etc. may go up. Similarly, benefits arising out of the project tend to be higher in future years for the same reason. Since relative costs and benefits remain almost the same, it is often common to disregard inflationary effects both in the cost and in the benefit streams. However, where these can be distinctly identified, these need to be considered in the analysis.

5.5.6.6   Limitations

Major limitations of shadow pricing technique are discussed in the following paragraphs:

a. Success of shadow pricing technique depends largely on availability of authentic data. However, this is often not easily available in developing or underdeveloped countries;

b. In developing or underdeveloped countries, it is not always possible to have a complete knowledge of demand and supply functions which are based on the existing socioeconomic environment in these countries. Thus, shadow prices are not easy to determine under the existing institutional structure;

c. Time dimension presents another problem. Shadow prices are generally used to overcome difficulties in the evaluation of projects which are basically progressive and not static. The anomaly arises because all inputs and outputs for shadow prices are valued at limited times, while investment projects relate to comparatively longer periods.

5.6 PREPARATION OF ECONOMIC APPRAISAL

In simple terms, the process for economic appraisal involves the following steps.

5.6.1 DEFINITION OF OBJECTIVE AND SCOPE

Objective and scope of the project must be defined very clearly and unambiguously at the very early stage.

5.6.2 IDENTIFICATION OF OPTIONS

Options that are realistic and at the same time far reaching are to be identified. This step should be taken as early as possible. Tendency to identify the solutions that have been attempted and discarded in the past should be avoided, as this can lead to dismissal of potentially successful option at the early stage without proper investigation.

5.6.3 IDENTIFICATION OF QUANTIFIABLE MONETARY COSTS AND BENEFITS

Economic appraisal should be based on all capital and recurrent monetary costs and benefits associated with a project. The degree of accuracy of the analysis depends on the authenticity of the source of the data.

5.6.4 CALCULATION OF QUANTIFIABLE NET BENEFITS

The quantifiable costs and benefits are expressed in net present value (NPV) terms using an agreed discount rate. By using discounting process, the operating costs and benefits, which may extend far into the future (may be 15–20 years), are brought back to a common time dimension – present value – for the purpose of comparison. (Basically, the process of discounting is a compound interest calculation worked backward.)

Based on the discounted stream of costs and benefits, the following decisions may be arrived at:

a. *Net present value (NPV)*: a project is potentially worthwhile if the NPV is greater than zero.
b. *Net present value per unit of investment (i.e., NPV/I)*: projects with the highest ratio of NPV/I would be potentially worthwhile.
c. *Benefit/cost ratio* (BCR): if the BCR is greater than 1, i.e., if the present value of benefits exceeds the present value of costs, the project is potentially worthwhile.

d. *Internal rate of return* (IRR): a project is worthwhile if IRR is greater than discount rate. (Internal rate of return (IRR) is the discount rate at which the NPV of a project is zero, i.e., discounted benefits equal discounted costs.)

(Note: For detailed discussions on NPV, BCR, etc., see Chapter 6.)

5.6.5   IDENTIFICATION OF QUALITATIVE FACTORS AND PREPARATION OF SUMMARY OF THE RESULTS

Apart from the quantifiable factors, there are qualitative factors that are also very important in the economic appraisal of a project and should be identified. These include environmental considerations, social and regional impacts, resource availability, marketing prospect, funding, distribution of benefits and costs, which also need to be taken into consideration.

5.7   CONCLUDING REMARKS

The basic philosophy behind economic appraisal is to maximize social benefits from spending of the public money. Although this technique is used mostly in public sector and public utilities projects, the same is often used for private sector projects also. There are instances that private sector organizations (nongovernmental organizations – NGOs) carry out corporate social responsibility (CSR) and promote healthcare units, educational institutions, etc. Economic appraisal of projects will be useful for such organizations also.

BIBLIOGRAPHY

Bhattacharjee, S.K., 2008, *Fundamentals of PERT/CPM and Project Management*, Khanna Publishers, New Delhi.

Chandra, P., 1996, *Projects: Planning, Analysis, Selection, Implementation and Review*, Tata McGraw-Hill Publishing Co. Ltd., New Delhi.

DeGarmo, E.P., Sullivan, W.G. and Canada, J.R., 1984, *Engineering Economy*, Macmillan Publishing Co., USA.

Dixon, R., 1994, *Investment Appraisal*, Kogan Page Ltd., London.

Gruneberg, S.L., 1997, *Construction Economics*, MacMillan Press Ltd., UK.

Kwat natasa, Shadow Prices: Meaning, Need, Limitations and Uses, http://www.economicsdiscussion.net (August 2019).

Manual on Economic Evaluation of Highway Projects in India, 1993, Indian Roads Congress, New Delhi.

White, J.A., Agee, M.H. and Case, K.E., 1989, *Principles of Engineering Economic Analysis*, John Wiley & Sons, Inc., USA.

6 Financial Evaluation

6.1 INTRODUCTION

Investment in a project is normally done with the expectation that any cash outlay now will result in extra cash (or benefits) in the future. Financial evaluation methods try to estimate whether the cash returns from the investment will be enough to justify the initial investment.

The following commonly used methods will be discussed in this chapter:

- Payback period;
- Return on investment (ROI);
- Net present value (NPV);
- Internal rate of return (IRR);
- Cost/benefit analysis (CBA).

6.2 YEARLY CASH FLOWS

For any project, estimates of initial cash inflows and outflows are necessary for understanding the pattern of cash expenses on certain items of the project. (The word 'cash' includes coins and notes as well as cheques, drafts and other banking instruments.) The fundamental concept of the topic is discussed in this section for the understanding of the financial evaluation techniques that follow. Typically, fixed assets such as buildings equipment, vehicles and raw material. are to be acquired at the beginning of the project. These would need an initial cash outflow. Depending on the necessity, additional working capital outflows may also be incurred at a later year during the duration of the project. Other cash outflows include operating (wages and raw materials) and maintenance costs. As the project is completed, the working capital is released, which is turned into cash inflow. In cash flow computation, the inflows (benefits) are treated as positive and the outflows (costs) as negative.

The main cash inflow in a project is from sales revenue, after deducting operating and maintenance costs. For easy understanding, the effect of inflation is ignored.

A typical computation of cash flow is shown in Table 6.1.

6.3 TIME VALUE OF MONEY (TVM)

Time value of money (TVM) is the concept that money has potential earning capacity (i.e., certain 'interest'), and therefore, money available at the present time (present value) is worth more than the same amount in a future time (future value). Future values of money are computed by using the concept of compound interest. Thus, if $10 is invested at 10% per annum and is left to accumulate interest, Table 6.2 shows how the capital grows in 3 years.

TABLE 6.1

Typical Cash Flow Computation over the Lifetime of a Project

Year	Annual Cash Flow ($)		Net Cash Flow ($)	Cumulative Cash Flow ($)
	Inflow	Outflow		
0	NIL	−13,000	−13,000	−13,000
1	NIL	−23,000	−23,000	−36,000
2	NIL	−16,000	−16,000	−52,000
3	+18,000	−5,000	+13,000	−39,000
4	+21,000	−2,000	+19,000	−20,000
5	+26,000	−5,000	+21,000	+1,000
6	+32,000	−6,000	+26,000	+27,000
7	+28,000	−2,000	+26,000	+53,000
8	+15,000	−2,000	+13,000	+66,000
9	---	−1,000	−1,000	+65,000

TABLE 6.2

Future Value of $10 at Compound Interest of 10%

Year	Future Value ($)
0	10.00
1	11.00
2	12.10
3	13.31

In short, compounding is the method of moving cash flows forward in time. Similarly, cash flows can be moved back in time also. In this case, money received or paid at a future time is to be treated as less value than the same amount received or paid today.

6.4 PAYBACK PERIOD

Payback period is generally used as an initial yardstick for reviewing and screening of any project option in terms of period of time that it takes to pay back an initial cash investment. Cash flows are accumulated annually and payback period is considered to have been reached when the cumulative cash flow reaches zero. In the cash flow, when there is initial outlay, the term 'Year 0' is used to signify the start of the project.

It should be noted that payback period uses cash flows only and not the net income. Also, it does not take care of the profitability of the project. It simply computes how fast the investment is recovered. Other parameters being equal, shorter payback periods are preferable to longer payback periods. Payback period is normally expressed in years. Table 6.3 illustrates a typical example of computation of payback period.

TABLE 6.3
Computation of Payback Period

Year	Annual Cash Flow ($)	Cumulative Cash Flow ($)
0	−40,000	−40,000
1	+12,000	−28,000
2	+14,000	−14,000
3	+8,000	−6,000
4	+8,000	+2,000

In this example, the payback period is between 3 and 4 years, and *the cumulative cash flow becomes positive in the 4th year.* In practice, the year in which the cumulative cash flow becomes positive is designated as the payback year.

6.4.1 ADVANTAGES

The basic advantages of payback period method are briefly stated below:

- Payback period is easy to comprehend and calculate. This is a 'rule of thumb' method for appraisal of projects of minor nature that come across frequently for investment decisions. Primary aim is to get the money back as early as possible so that it can be re-invested in other projects. Thus, it is a useful capital budgeting method for cash-starved business organizations;
- It tries to take care of the risk factor in an investment. In cases where risk is anticipated from political, economic or any other direction, the shortest payback period should be the best option as the shortest time period reduces the risk of unforeseen happenings.

6.4.2 DISADVANTAGES

In spite of its simplicity to calculate, the payback period method has some inherent disadvantages. These are briefly discussed below:

- Payback period does not take into account the TVM and consequently does not present the true financial scenario while evaluating the cash flows of different options of a project. The issues related to TVM (e.g., NPV, IRR and discounted cash flow) will be discussed in later sections;
- Payback period is an arbitrary period without any rationale for selecting a particular project. It provides primary emphasis on the shortest period of time for returning the investment and ignores cash inflows after the payback period. Thus, the method is likely to reject an alternative and perhaps a better investment proposal simply because it does not meet an arbitrary payback period target. This, certainly, is not a sound financial decision. The weakness is illustrated by an example. Two investment options are compared, viz., Project A and Project B, each requiring an investment of $ 40,000. The expected cash flows are shown in Table 6.4.

TABLE 6.4

Comparison of Investment Options

| Year | Project A | | Project B | |
	Annual Cash Flow ($)	Cumulative Cash Flow ($)	Annual Cash Flow ($)	Cumulative Cash Flow ($)
0	−40,000	−40,000	−40,000	−40,000
1	+12,000	−28,000	+4,000	−36,000
2	+14,000	−14,000	+6,000	−30,000
3	+8,000	−6,000	+10,000	−20,000
4	+8,000	+2,000	+12,000	−8,000
5	+2,000	+4,000	+12,000	+4,000
6			+10,000	+14,000
7			+10,000	+24,000

In this example, as per payback period method, Project A takes lesser period for recouping the initial investment and should be accepted. However, this would mean rejection of a more profitable option, viz., Project B, which continues to give revenues for more years after the payback period of the alternative option (Project A);

- Payback period method ignores the timing and the quantum of the cash flows within the payback period. In Table 6.5, two identical investment proposals, both having the same payback period, are compared. Cash inflow in case of Project C is more in the first two years compared to that in the case of Project D, and consequently, Project C has an advantage over Project D. But at the same time, it has to be considered that cash inflows after the payback period are much higher in the case of Project D. This aspect is required to be considered during selection process.

TABLE 6.5

Comparison of Investment Options

| Year | Project C | | Project D | |
	Annual Cash Flow ($)	Cumulative Cash Flow ($)	Annual Cash Flow ($)	Cumulative Cash Flow ($)
0	−60,000	−60,000	−60,000	−60,000
1	+20,000	−40,000	+10,000	−50,000
2	+20,000	−20,000	+10,000	−40,000
3	+10,000	−10,000	+20,000	−20,000
4	+10,000	NIL	+20,000	NIL
5 (Onwards)	+10,000	+10,000	+20,000	+20,000

6.5   RETURN ON INVESTMENT (ROI)

This method is also known in a number of titles, of which accounting rate of return (ARR) and return on capital employed (ROCE) are some of the most commonly used titles. In this method, an average rate of return is calculated by expressing average annual profit as a percentage of average capital investment in the project. Thus,

$$\text{ROI} = \frac{\text{Estimated average annual profit}}{\text{Average capital invested}} \times 100$$

This can be illustrated by an example.

Rs. 1,70,000/- is invested in a project, and the estimated annual profits are as follows:

Year 1: Rs. 10,000/-
Year 2: Rs. 20,000/-
Year 3: Rs. 20,000/-
Year 4: Rs. 10,000/-
Total profit = Rs. 60,000/-

$$\text{Average annual profit} = \frac{\text{Rs.}60{,}000}{4} = \text{Rs.}15{,}000$$

$$\text{Average capital invested} = \frac{\text{Rs.}1{,}70{,}000}{2} = \text{Rs.}85{,}000$$

$$\text{Therefore, ROI} = \frac{\text{Rs.}15{,}000}{\text{Rs.}85{,}000} \times 100$$

$$= 17.65\%$$

6.6   NET PRESENT VALUE (NPV)

In this method, all future costs and benefits related to a project are brought together into a single value by applying the traditional 'present-value' concept on a base date corresponding to the initial costs. The inflows (benefits) are treated as positive and outflows costs as negative, and the summation gives the NPV.

Thus:

$$\text{NPV} = \sum_{t=1\,\text{to}\,n} \frac{(b-c)}{(1+r)^t}$$

where
b = Benefits
c = Costs

r = Selected discount rate per annum expressed in decimal
t = Time in years when the future cost is incurred
n = Number of years considered in the analysis (life of the project)
The following example will make the matter clear:

A discount rate of 10% is assumed for a project with an initial outlay of $10,000 in year 'one'. In year 'two', the net income is $7,000 and again $6,000 in year 'three'. The NPV will be calculated as follows:

$$NPV = -10,000 + \frac{7,000}{1.10} + \frac{6,000}{(1.10)^2}$$

$$= +1,322$$

A positive NPV implies that estimated total benefits exceed total costs. While comparing alternative proposals, the project with higher NPV is to be preferred, other factors being equal.

6.7   INTERNAL RATE OF RETURN (IRR)

The IRR of a project is the rate of return (discount rate) which equates the discounted net benefits to discounted capital costs, and can be obtained by solving for r in the following equation:

$$C_0 = \sum_{t=1\,to\,n} \frac{(b-c)}{(1+r)^t}$$

where
    C_0 = Initial outlay
    b = Benefits
    c = Costs
    r = Selected discount rate per annum expressed in decimal
    t = Time in years when the future cost is incurred
    n = Number of years considered in the analysis (life of the project)

As an illustration of computation of IRR, consider that the initial outlay for a project is $10,000 and the subsequent annual returns are $3,000, $4,000 and $6,000 in years one, two and three at a return rate of r. This can be represented by the following equation:

$$10,000 = \frac{3,000}{(1+r)^1} + \frac{4,000}{(1+r)^2} + \frac{6,000}{(1+r)^3}$$

Solving the above equation for IRR of a project is rather tedious and needs a trial-and-error method if done manually. This can be solved easily by using a computer.

If the IRR calculated from the above formula exceeds the rate of interest obtained by investing capital in the open market, the scheme may be considered acceptable.

6.8   COMPARISON BETWEEN NPV AND IRR METHODS

In NPV method, the discount rate is assumed first and the NPV is calculated accordingly. However, in the IRR method, the present value is set to be equal to zero, and then the 'rate' (IRR) that satisfies this condition is determined. Essentially, IRR does not produce any new information compared to NPV. It indicates the same information in a different manner.

6.9   BENEFIT/COST RATIO (BCR)

In this method, all benefits and costs are discounted to their present value, and the ratio of the benefits to costs is calculated. As in the previous cases, benefits and costs are considered as positive and negative values, respectively.

Thus:

$$BCR = \frac{\text{Discounted benefits}}{\text{Discounted costs}} = \frac{\text{NPV}}{C_0}$$

where the symbols are as before.

A project with a BCR greater than one is worth undertaking. Also, for comparing purpose, higher the value of BCR for an alternative, the higher will be its ranking.

6.10   DRAWBACKS OF BCR

BCR method is a popular method used by the analysts. However, it suffers from a few drawbacks. The major drawback is that the discount rate needs to be assumed. This rate is related to the opportunity cost of capital, which is not easy to ascertain.

6.11   CONCLUDING REMARKS

The methods discussed in this chapter are extensively used for financial evaluation of projects. These involve four basic parameters, viz., initial costs, future costs, predicted time frame and the discount rate. While initial costs can be reasonably estimated, there are some misgivings about the accuracy in forecasting the other parameters. Although these perceived difficulties are real to varying degrees, the method is often utilized as a very useful qualitative tool for decision making. It addresses the future costs in proper perspective and offers an informed and rational way of evaluation of alternative solutions. It also brings to focus that the future costs are required to be given due importance in the early stage of the project itself and that the concerned stakeholders should be aware of the consequences of their present actions. To make it more effective, reliable inputs such as database of future costs, service life, as well as appropriate discount rates in respect of different alternatives, are to be made available in the early stage of the project cycle.

BIBLIOGRAPHY

Abelson, P., 1996, *Project Appraisal and Valuation of the Environment*, Macmillan, London.

Bhattacharjee, S.K., 2004, *Construction Management of Industrial Projects*, Khanna Publishers, India.

Chandra, P., 1996, *Projects: Planning, Analysis, Selection, Implementation, and Review*, Tata McGraw-Hill Publishing, New Delhi.

Dixon, R., 1994, *Investment Appraisal*, Kogan Page, London.

Fraser-Sampson, G., 2011, *No Fear Finance*, Kogan Page, London.

Gruneberg, S.L., 1997, *Construction Economics*, Macmillan, London.

Mott, G., 1992, *Investment Appraisal*, Pitman Publishing, London.

Oldcorn, R. and Parker, D., 1996, *The Strategic Investment Decision*, Pitman Publishing, London.

Robson, A.P., 1997, *Essential Accounting for Managers*, Cassel, London.

Warren, M., 1993, *Economics for the Built Environment*, Butterworth-Heinemann, Oxford, UK.

7 Environmental Appraisal

7.1 INTRODUCTION

Environmental appraisal of a project is a process for identifying and evaluating the potential benefits as well as adverse impacts of a project on the surrounding environment. It provides a clear, well-structured and rational analysis of the consequences of proposed actions, and assists in selecting the most environment-friendly option. It also helps in reducing the project cost associated with avoidable subsequent delays.

The insatiable demands of modern society on our environment have exceeded dangerously since the 1970s. As a result, the planet's resources are being exhausted posing threats to our health and welfare. It is high time that environmental analysis is given appropriate importance prior to the selection of a project for implementation.

The process of environmental analysis originated in the United States of America in the late 1960s and the early 1970s, and has since been used increasingly around the world. Initially, it was largely adopted in a few high-income countries like Canada, Australia and New Zealand (1971–1974), followed by some developing countries like Columbia (1974) and Philippines (1978). As awareness among more people increased, the issue became a common topic for debate and discussion. Consequently, the average person became increasingly more concerned about the impact of environment. In the 1980s, World Bank and other international agencies introduced environmental requirements for compliance in all projects. Gradually in most countries (including less developed countries), these became important considerations while evaluating the feasibility of a project along with technical and financial viability. Currently, environmental analysis is an accepted procedure for utilizing the natural resources available within the system. It is now a prerequisite for funding by international agencies like World Bank, United Nations Industrial Development Organization and Asian Development Bank. It is normally carried out if a project is likely to have a significant impact on environment because of its *prima facie* nature.

7.2 PROJECT TYPES

Projects may be divided broadly into two types, viz., production-oriented and service-oriented. The former category includes chemical plants, metal industry, refineries and cement plants. These production-oriented projects are involved in transforming natural resources to saleable goods and have direct impact on the environmental and ecological balance. The second type of projects involve rendering various services, such as education, health, law, defense and land reforms. These categories of projects do not have an immediate impact on the environment. However, these may create far-reaching outcome in the future on values, lifestyles and social links leading to promoting consumerism in the society and consequently indirectly encouraging manufacturing projects. Thus, there is a link between the two types of projects.

7.3   MEANING AND SCOPE OF ENVIRONMENT

According to Concise Oxford Dictionary, the word 'environment' means 'the sur-roundings or conditions in which a person, animal or plant lives or operates'. In other words, it comprises the land, air and water bodies in the neighborhood of the project, and covers all items that inhabit in the surroundings, as well as the social, economic and political systems in which people function. World Bank's broad defini-tion of environment as 'The natural and social conditions surrounding all mankind, including future generations' is pertinent in this context (World Bank, 1992, quoted by Peter Abelson).

Basically, environment may be classified into two categories, viz., organic and inorganic. First, the organic setting consists of animate elements such as human beings, animals, plants and also other living organisms like bacteria and viruses. The second category comprises inorganic elements like land, water and atmosphere.

The environment boundary stretches beyond the legal boundary of a project and covers the areas which are likely to be 'environmentally' affected by various natural factors such as direction and speed of wind and elevation. Thus, it varies from project to project as well as from location to location.

7.4   MAIN ENVIRONMENTAL ISSUES

The aspects concerning impact of a project on the environment which should be con-sidered prior to discussion of the main issues may be broadly identified as follows:

- The existing environmental and socioeconomic conditions of the site;
- Effects of the proposed project on these conditions;
- Examination of the impact of the proposed project *vis-à-vis* the existing environmental regulations. Any impact that exceeds the regulations should be eliminated to avoid environmental hazard.

The main issues concerning environmental analysis are discussed in the sections that follow.

7.4.1   POPULATION GROWTH

During the past 50 years, earth's population has grown at an unprecedented rate. As a result, there has been enormous pressure on both renewable and nonrenew-able resources. This has resulted in a reduction in the amount of investment as well as productivity per worker, leading to an increase in the inequality of income and also crowding and congestion. According to World Bank's prediction, global popula-tion will double to around 10.5 billion by 2050, and a vast majority of this growth will be in the developing countries. Also, urban population (which currently exceeds the rural population) will double itself by 2030. As a result, the demand for food is expected to increase threefold. Simultaneously demand for goods and services will also rise enormously. Also, the increase of waste material will be an added burden to the waste disposal system.

7.4.2 Global Atmospheric Changes

During the past two decades of the 20th century, there have been instances of atmospheric changes in many parts of the world, such as depletion of ozone concentration and increase in carbon dioxide emission causing global warming (greenhouse effect).

It is difficult to accurately quantify ozone concentration in the atmosphere. However, it appears that ozone has depleted dramatically over large parts of the world causing a hole in the ozone layer above Antarctica. The reduced ozone is likely to increase ultraviolet radiation reaching the earth, and cause health hazards to human race (e.g., skin cancer, eye damage, depletion of immune response system)

Carbon dioxide and other greenhouse gases increase the atmospheric temperature considerably. Global warming is likely to affect low-lying coastal properties adversely (by flooding), agricultural productivity, water supply, electricity requirements, etc. Other examples of damage are disruption in marine food chain, threat to biodiversity and ecological impacts.

7.4.3 Pollution

Pollution is the environmental degradation caused by the introduction of harmful substances into the natural environment beyond a level that causes instability, harm or discomfort to the systems living therein. These can be naturally occurring substances, or energy such as noise, heat or light. These are considered as contaminants when these are in excess of natural levels. The major forms of pollution are briefly discussed in the following paragraphs.

7.4.3.1 Air Pollution

Air pollution has a serious impact on health, particularly in urban areas where risks of respiratory disorders are common. In such areas, high levels of sulfur dioxide and lead substantially reduce the intelligent quotient (IQ) level of children and increase hypertension disorders, including heart attacks and strokes in adults. These pollutants emanate mostly from traffic, industrial plants and power-generating plants. Dust, smoke and other dry particles also damage the cleanliness of the sky resulting in fog, mist, haze, smoke, smog, etc. Smog causes headaches, irritations in the eyes, nose and throat and reduced visibility, and can indirectly affect lung function also. Air pollution may also lead to damage to crops and corrosion to buildings.

Another aspect of air pollution is 'acid rain' or 'acid precipitation'. These terms are commonly used to mean deposition of acid components in rain, snow, fog, dew, etc. Acid rain originates from sulfur and nitrogen oxides from power stations and motor vehicles. Generally, clouds carry these pollutants and deposit these in rain and snow. Acid rain causes considerable harm to natural areas like lakes, rivers, forests, as well as to crops and fish population.

Indoor air in workplace is sometimes more polluted than outdoor air. Often smoky indoor air causes greater health hazard than the outside environment. Most indoor pollution is caused by use of coal, wood, etc. These produce pollutants like sulfur dioxide, carbon monoxide and lead. There are instances when indoor air pollution remains undetected and unreported causing sickness or even death. Appropriate ventilation required for diluting the contaminants, filtration and control of the source could improve air quality in most dwellings.

7.4.3.2   Water Pollution

In many parts of the world, including India, China and Australia, the demand for pollution-free (clean) water resources greatly exceeds the supply. As a result, a large number of people are forced to consume unclean water, thereby exposing themselves to various kinds of diseases, such as diarrhea, typhoid, paratyphoid and cholera.

Water pollution is largely caused by discharge of wastewater from industrial or commercial waste into surface waters, discharge of untreated chemical contaminants (e.g., fertilizers and pesticides), domestic sewage, etc. Urban runoff and agricultural runoff (which may contain chemical fertilizers and pesticides), waste disposal and removal of specific soluble substance from water are other causes of water pollution. Also, marine oil spill, i.e., release of liquid crude oil or petroleum product into the ocean or coastal waters, may create heavy water pollution. It may take several months to clean up the degradation. Oil spill often occurs in marine environment, where oil is released into the ocean or coastal waters. Oil may be crude oil, petrol or diesel. Man-made oil pollution mostly originates from land-based activities, although public focus is mostly on sea-going oil tankers. A type of water pollution is termed *eutrophication*. This is a natural process in which inland water bodies receive plant nutrients as a result of natural erosion and draining away of water from surrounding land. *Hypoxia* or oxygen depletion is another type of water pollution, in which the dissolved oxygen in the water is reduced in concentration causing detriment to the aquatic organisms living in the system. Apart from health hazard, water pollution and scarcity is likely to restrict economic output seriously and cause decline of fishery industry, boating, swimming, etc. It also will make safe water costlier.

One other aspect needs special attention. Risks of water pollution and costs for avoiding it may increase because of deterioration of available infrastructure.

7.4.3.3   Solid and Hazardous Wastes

Environmental degradation due to solid and hazardous wastes presents a massive problem worldwide.

As regards solid wastes, in many developing countries, industries and households receive waste disposal service. Landfilling, composting and recycling plants and incinerators are the usual methods of disposal. But unfortunately, that is not the case in many developing countries. Quite often, there are no official waste disposal services and solid wastes are randomly dumped at informal sites, causing major health risks due to breeding of rats, flies and mosquitos. These cause air pollution and also water pollution by leaching of chemicals into the surface and groundwater system. Thus, routine disposal of solid wastes is likely to cause water or air pollution. Also, dumping of solid wastes in waterways blocks drains and causes flooding.

Hazardous wastes comprise toxic, infections and corrosive wastes, and cause major health risk to the society. These wastes originate primarily from mining, petroleum, metallurgical industries, and tanneries, car battery recycling outlets, etc. Routine nuclear waste is another hazard that has a huge geographical impact. With increasing number of nuclear reactors, nuclear weapons research, manufacture and deployment worldwide, this hazard carries a chronic risk. Continuous disposal of hazardous waste may cause severe water and air pollution. Furthermore, instances of industrial and nuclear radiation accidents are very much there. Examples are Bhopal, India, and Chernobyl, Russia, where large-scale catastrophes took place causing loss of life and other serious damages.

7.4.3.4 Noise Pollution

Globally, millions of people live in crowded conditions and are exposed to unsatisfactory acoustic environment. The condition has been deteriorating during the past three decades due to increase of road and air traffic. Thus, in modern society, noise pollution is an inbred phenomenon and is very difficult to get rid of.

Noise pollution is measured in decibel (dB), a logarithmic unit of measurement of sound intensity relative to a reference level. Since this is a ratio of two same-unit quantities, it is a dimensionless unit. Excessive loud or high-pitched sound can cause discomfort and damage the hearing capacity of the ear. Thus, people living near airports or working in noisy industrial units may suffer from these effects of noise pollution. Also, the crew members of public transport system are often affected by prolonged exposure to honking and other damaging noise pollution. This may impair the hearing capacity of the driver and lead to future road accidents because of his failure to hear horns.

7.4.3.5 Other forms of Pollution

Other forms of pollution include the following:

- *Thermal pollution*: temperature change in natural water bodies for mainly industrial purpose, e.g., use of water as cooling agent in power plants (cooling towers).
- *Visual pollution*: highway advertisement boards, overhead cable lines, open storage of trash, municipal solid waste, etc.
- *Light pollution*: highway luminal advertisements etc.
- *Plastic pollution*: stock piling of plastic products in the neighborhood that adversely affects wildlife, as well as human society.

7.4.4 SOILS AND FORESTS

Depletion of soil productivity is caused generally by nutrient deficiency, soil erosion, desertification, water pollution, salinization, etc. Increased use of fertilizers may increase soil productivity, but that is only a temporary solution. Soil productivity loss has decreased agricultural productivity in many countries. Salt deposited from irrigation water is another problem affecting soil productivity. In order to maximize output from limited farming lands, extensive use of fertilizers and pesticides is often resorted to. This, in turn, exhausts soil nutrients and affects soil productivity adversely. This also pollutes waterways.

Desertification is a major problem in sub-Saharan Africa and China where productive semi-arid areas are being turned into deserts. This phenomenon occurs due to natural droughts, overgrazing and soil erosion. Also, there are instances where desertification has been caused not only by natural droughts, but also by soil erosion.

Deforestation activity is often undertaken in order to increase agricultural productive land. This leads to reduction of forest products and direct income of the affected community. On the other hand, deforestation increases erosion of the soil, especially in the hilly arrears, and adversely affects ecological balance with impacts on regional and global climates. It is therefore imperative that deforestation is carried out systematically so that forest resources are increased without hampering the ecological balance.

7.5   CONCLUDING REMARKS

In the foregoing sections, several instances of possible environmental degradation due to new projects have been discussed. However, it should be understood that in parallel with these disruptive effects, quite often projects have positive impacts on the society. These may help to introduce new resources by inducting new technologies in the field and thereby utilize skill of people. For example, a river valley project may cause flooding of vast areas of habitation and agricultural land, but certainly helps in irrigation and generation of hydroelectric power, thereby indirectly helping industrialization of the area and its socioeconomic development in the long run.

The other aspect that needs to be mentioned is uncertainty and lack of knowledge about the consequences of environmental degradation. The impact of air and water pollution on public health, of acid rain on crop productivity, of greenhouse gases on global warming and of deforestation on climate change, to name a few examples, is clouded with uncertainty. Thus, it is very difficult to make precise estimates of the impact of these examples of environmental degradation on the health and productivity of the community.

BIBLIOGRAPHY

Abelson, P., 1996, *Project Appraisal and Valuation of the Environment*, Macmillan Press Ltd., London.

Chandra, P., 1996, *Projects: Planning, Analysis, Selection, Implementation, and Review*, Tata McGraw-Hill Publishing, New Delhi.

Investment Project Preparation and Appraisal (IPPA) Teaching Materials (Module 3), 2005, United Nations Industrial Development Organization, Vienna.

Lahiri, D., 2009, *Environment: The first child of Nature*, Sahitya Samsad, Kolkata.

Spellman, F.R., 2010, *The Science of Environmental Pollution*, CRC Press, Taylor & Francis Group, USA.

8 Uncertainty and Risk Analysis

8.1 INTRODUCTION

Initial estimated costs and financial benefits of a project are based on assumptions of expenses and revenues of some quantifiable variables, the values of which are based on the data available at the time of the budget. However, values of these variables are often at variance from the subsequent actual figures. This is mostly due to occurrence of some unexpected events (uncertainty and consequent risks). Possibility of such uncertainty and risk does exist in every project. The various types of uncertainty and risks are illustrated by the following examples:

- Sales of goods or services not as per expectation;
- Costs of goods or services higher or lower than budgeted figures;
- Variation in business environment (change in interest rate, tax structures inflation, government policy, etc.);
- Changes in technologies or project life cycle;
- Industrial relations problems;
- Unavailability of resources.

All uncertainties lead to risks, which may affect the success of the project in terms of:

- Budget;
- Completion date;
- Scope;
- Quality.

Project risk analysis is aimed at removing/reducing the risks that threaten attainment of the success of the project.

8.2 UNCERTAINTY AND RISK

In financial and economic analysis, the word 'uncertainty' is referred to as 'unreliability' of actual values *vis-à-vis* estimated values. The causes of unreliability may be due to errors in estimating due to insufficient information, inability to predict the future, etc. The word 'risk' means volatility of expected returns. Volatility conveys the idea of change from anticipated (estimated) values, which may be due to unplanned causes. Actual values can be known only on completion of a project. Some authorities consider *risk is uncertainty that matters*. Thus, uncertainty appears to be a generic term, while risk is a more specific term. In any case although technical distinction may exist between the words uncertainty and risk, there is hardly any significant gain for the present text in treating these two words as different. Therefore, these two words are intended to be used synonymously and interchangeably in the present text.

8.3　BENEFITS OF RISK ANALYSIS

Benefits of risk analysis are manifold. The foremost benefit is a better understanding of the project situation leading to appropriate future planning in respect of time frame and costs. It also helps in efficient and effective management of the risk. This results in discouraging acceptance of financially unsound projects. One other use of risk analysis is a better understanding of the risk and, if necessary, the most suitable agency may be engaged to tackle it. Better understanding of the risk may also lead to the incorporation of appropriate clauses in the contract agreement for the project. Risk analysis should be properly documented since the resulting data may be useful for future projects.

8.4　FACTORS CONTRIBUTING TO RISK

There are several factors that may cause uncertainty and risk in a project. Some of the major factors are discussed below.

8.4.1　INACCURACY IN THE INPUT DATA

The first key factor for risks in the analysis is the possibility of inaccuracy in the input data which form the basis of the study. Thus, the level of reliability of the study depends largely on the accuracy of the information available. In cases where the information is based on past experience or adequate market surveys, this may be fairly dependable. On the other hand, if the information is based on mere guesswork, a fair amount of uncertainty is likely to creep in.

8.4.2　TYPE OF THE ACTIVITY AND FUTURE TREND OF ECONOMY

The second factor causing uncertainty is the type of activity involved in the project and the future trend of the economy. While making capital investment in any enterprise, it is imperative to study in advance the nature and background of the enterprise as well as the projected future economic conditions (e.g., interest rates, inflation) and assess the risk involved in the process. However, this may not be easily achievable, since for new enterprises, hardly any past history is available, particularly in the current fast-changing business cycles.

8.4.3　TYPE OF PLANTS AND EQUIPMENT

A third factor causing uncertainty and risk is the type of plants and equipment involved in the project. Some plants have definite economic lives and resale value, while others may not have much resale value. This will have direct bearing on the anticipated income and expenditure. Multipurpose equipment may be very useful in the workshop, while equipment capable of only limited specialized job would naturally have restricted use. This aspect needs careful examination while studying the investment strategy.

8.4.4 EXTENT OF THE ESTIMATED STUDY PERIOD

If the extent of the estimated study period is too long, it may not be possible to maintain the income/expenditure figures as per the initial estimation. Therefore, a long study period will increase the probability of uncertainty and risk in an investment.

8.5 IDENTIFICATION OF RISKS

It is vital that project risks are identified before the project is implemented, preferably at the appraisal stage itself. This would help in developing appropriate strategy for managing such risks. In this connection, past experience would be the ideal way of identifying the probable risks in the project.

Past experience may be of two categories. First, personal experience of the members of the project needs to be explored. Dialogue with team members seeking their inputs from personal knowledge within their specific field of expertise may help the team in this pursuit. Second, records of past experience of the organization may help in identifying future risks of similar nature and prepare for and/or mitigate the damage.

8.6 ANALYSIS OF RISKS

Analysis of risks is a systematic process for the assessment of identified risks in a project. There are several methods for analyzing risks. Some of the methods are detailed in the succeeding sections.

8.6.1 QUALITATIVE RISK ANALYSIS

Qualitative method of risk analysis is generally used to prioritize the already identified risks by utilizing the data in respect of their impact on the project objectives and the rate of their occurrences. It is often useful for making an estimate of the extent of the risk and can provide useful qualitative assessments to stakeholders. Also, compared to quantitative analysis, this method (qualitative analysis) is often more accessible and more easily understood by the stakeholders. In particular, where sufficient data or mathematical competence and facilities for assessment are not available, qualitative analysis becomes quite a useful and handy method of risk assessment.

8.6.1.1 Probability and Impact Matrix

This method is one of the common methods of qualitative risk assessment. The method is used to evaluate the importance and to prioritize each risk. Risk can be ranked in respect of each of the project objectives. For the purpose of general ranking, the advantage among the several project rankings can be assessed without difficulty. This makes the process easily workable.

TABLE 8.1
Probability and Impact Matrix of Risks

		Impact Severity		
		Low (Insignificant)	**Medium** (Reasonable)	**High** (Significant)
Probability of occurrence	Low (unlikely to occur)	5	4	3
	Medium (may occur)	4	3	2
	High (likely to occur)	3	2	1

Actual technique of this method is to allot values for two items.

- Probability of occurrence of each risk;
- Impact of the risks.

Ranges of these values are to be assessed by the project team and stated in a matrix form. A typical probability and impact matrix is illustrated in Table 8.1. In this case, the risk grading is considered in descending order of risk, i.e., '1' is considered to be of highest risk and '5' is considered to be of lowest risk.

8.6.1.2   Brainstorming

Brainstorming may be defined as 'a means of getting a large number of ideas from a group of people in a short time'. It is a simple, yet effective, way to think creatively in a group setting without any emotional blockage or criticism from other members.
   The definition embodies three aspects:

- A large number of ideas;
- A group of people;
- A short time.

The intention of brainstorming is to allow all the stakeholders to use their knowledge and experience, and contribute effectively in the generation of ideas for risk analysis within a reasonable time frame. This method, however, cannot provide quantitative results, which is one of its drawbacks.

8.6.1.3   Delphi Method

The basic philosophy of Delphi method is to bring out the consensus of a panel of experts with the help of a mail survey. This method is often used for assessing potential risks in activities where the impact of risk is of prime importance.
   First, the group leader selects a panel of experts on the topic being examined. This activity is of great importance as the result of the study would depend on the level of expertise of each and every participant. Once the expert participants are selected, each member of the group is sent a questionnaire by mail and asked to comment on each item based on their personal opinion, experience or previous research. The feedbacks received from the experts are summarized without divulging the identity of the

experts. A copy of the summary of the feedbacks is sent to each participant, for his further comments. The questionnaires are returned to the group leader who prepares a summary and sends this again to the participants for their comments, if any. The questionnaire rounds are repeated till a reasonable consensus emerges.

The advantage of this method is that it attempts to elicit opinions from relevant experts without having to bring them together in a physical meeting. Also, since the names of the experts are kept anonymous, individual panelists are kept free from worrying about repercussions for their opinions. Furthermore, it is less expensive than the conventional face-to-face meetings. While Delphi method is appealing, as this method allows consensus of a diverse panel of experts, it has its disadvantages too. First, it does not provide the same sort of live interactions, which is the essence of traditional face-to-face meetings. Also, response time may be long, and the method is time-consuming. The other disadvantage is that the information received back from the expert panelists may not provide any intrinsic value.

8.6.1.4 Lessons Learnt from Previous Projects

Final reports of previous projects may reveal experiences gathered by earlier project teams in respect of certain risks. The present project team would then save both time and money if the previous experience is utilized in avoiding repetition of the same mistake.

8.6.2 Quantitative Risk Analysis: Sensitivity Analysis

Compared to qualitative risk analysis, quantitative risk analysis includes more sophisticated methods. Currently, computer software is available for such analysis. Quantitative analysis is effectively a continuation of the qualitative analysis. Once the risks are identified by qualitative analysis, a detailed quantitative analysis may be carried out. With this exercise, the impact of the risk may be quantified in respect of the three basic project success yardsticks, viz., cost, time and performance. Several methods have been developed for this purpose, among which sensitivity analysis is the simplest method. It involves changing the value of a few selected variables and calculating the resulting change in the net present value (NPV). In practice, a number of variables are responsible for the economic viability of a project, viz., economic life of the project, change in the interest rate of the borrowed capital, selling price, etc. Any change in these basic parameters will automatically change the NPV of the project. Sensitivity analysis shows the effect of such variations upon the expected return to be achieved in a project. For carrying out sensitivity analysis, the estimated values of the basic variables are considered, taking one factor at a time. The sum total of NPV of the project is then calculated. The effect of the NPV is plotted graphically for examination.

NPV of a project is represented by:

$$NPV = \sum_{n=1}^{n} \frac{Q(P-V)-F}{(1+i)^n} + \frac{S}{(1+i)^n} - I$$

where
    n = expected project life in years
    Q = number of units sold annually

P = selling price per unit
V = variable cost per unit
F = fixed costs, including taxes and liabilities
i = interest rate
S = salvage value
I = initial investment

It may be noted that the basic variables in the above equation are represented by selling price (P), interest rate (i) and the expected project life (n). From the above equation, an idea in the change of NPV can be formed, when one of the variables, say the selling price (P), is changed, keeping the other variables unchanged. Similarly, an idea of the change in the NPV can be obtained, if the interest rate (i) is varied, keeping the other variables unchanged. Figures 8.1 and 8.2 illustrate the visual representation of typical changes in NPV if the selling price (P) or the interest rate (i) is varied. Such visual representations are helpful in identifying variables that are critical to the success of a project. It may be noted from Figures 8.1 and 8.2 that when the selling price is too low or the interest rate is too high, the value of NPV indicates unfavorable investment condition.

Apart from being one of the most appropriate methods for indicating favorable and unfavorable investment condition for appraisal of projects, sensitivity analysis can also help in the initial stages in the design process of a project by:

- Improving the understanding of the effect of change in variable factor(s) on the project;
- Increasing expected NPV by varying the design in the initial stage;
- Reducing risk factor by taking adequate precaution;
- Indicating areas that need more in-depth investigation.

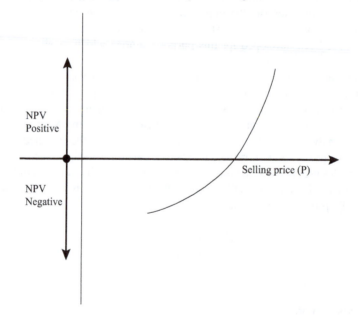

FIGURE 8.1   Relation between P and NPV.

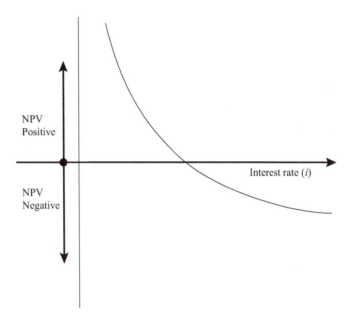

FIGURE 8.2   Relation between *i* and NPV.

Sensitive analysis is thus not only useful, but also very simple and consequently very popular in analyzing the risk in an investment proposal. However, when more than one variable changes simultaneously at a time, the analysis may not give an authentic representation.

8.7   CONCLUDING REMARKS

The information collected during risk analysis phase is generally used to improve the probability of the project in achieving the targets of its cost, time and performance objectives. This is done by reducing the risk whenever possible, and monitoring the risks where reduction is not possible.

This may involve changing the project schedule by removing high-risk activities or setting up monitoring device for getting early responses to risks. These responses are as follows:

• Eliminating risks whenever possible;
• Decreasing the risks by taking appropriate actions immediately;
• Mitigating the risk by contingency.

The risk management phase begins as soon as the qualitative analysis phase is complete, and proceeds through the quantitative analysis phase (sensitivity analysis). During this phase, the management may take action against the impact of the risk, say by changing the overall time frame or cost plan in order to increase the probability of achieving his project objective.

BIBLIOGRAPHY

Bhattacharjee, S.K., 2008, *Fundamentals of PERT/CPM and Project Management*, Khanna Publishers, New Delhi.

De Garmo, E.P., Sullivan, W.G. and Canada, I.R., 1984, *Engineering Economy*, Macmillan Inc., USA.

Hillson, D., 2009, *Managing Risk in Projects*, Gower Publishing Ltd., UK.

Oldcorn, R. and Parker, D., 1996, *Strategic Investment Decision*, Pitman Publishing, London.

Rawlinson, J.G., 1994, *Creative Thinking and Brainstorming*, Gower Publishing Co. Ltd., UK.

White, J.A., Agee, M.H. and Case, K., 1989, *Principles of Engineering Economic Analysis*, John Wiley & Sons, USA.

9 Life Cycle Costing

9.1 INTRODUCTION

When a project is examined for implementation, many issues such as user benefits, environmental aspects, defense considerations, political necessities, financial considerations and economic considerations come up for consideration. However, since resources are scarce and are required to be expended judiciously, cost consideration becomes a significant element among these various issues. The approach towards arriving at the optimum cost has changed over the years. The traditional practice for financial evaluation of any project is to base the decision solely on the initial capital costs of different options with little regard to future operating and maintenance costs or replacement intervals. It is, however, increasingly being felt that the evaluation should be done on a suitable basis, which considers the long-term effects of different options. Life cycle costing (LCC – sometimes referred to as whole life costing: WLC), can fulfill this objective. The philosophy behind LCC is that not only the initial costs, but also all future operating and maintenance costs to be incurred during the subsequent life spans of the competing projects, are to be considered for evaluating the total costs. Thus, in an aggressive environment, a special corrosion-resistant painting system, which would initially cost more, but would last longer than normal painting system may prove to be more economic over a period of time if analyzed with LCC. Additionally, the life of the structure may be extended, thereby delaying heavy capital expenditure for replacement of the structure.

In this system, the future costs are brought together into a single value, often by applying the traditional present value concept on a base date corresponding to the initial costs. It is thus useful to evaluate trade-off situations, such as choosing between one asset and another or choosing a higher initial cost item to save future running costs.

9.2 KEY STAGES OF A PROJECT

Life cycle of a project is the time period between conception of an idea about an asset and its final disposal.

The key stages comprise:

- Conception of the idea;
- Strategic planning;
- Creation of the asset;
- Operation;
- Maintenance;
- Rehabilitation;
- Decommissioning;
- Disposal.

9.3   BASIC CONCEPT OF LIFE CYCLE COSTING

The philosophy behind LCC is that not only the initial costs of the asset but also the future costs incurred during the subsequent life span of the asset are put into a comparable form and brought together into a single value for the purpose of comparison with alternative solutions. Therefore, by applying LCC concept, a solution with a higher initial cost may be justified because of reduced future running costs compared to a solution with a lower initial cost, but with high running costs over the study period. Thus, the lowest initial cost of a project is not necessarily the lowest total cost, if we consider LCC. While calculating the future costs, the benefits or profits from the project need also to be taken into account. The method can be used for comparison of different solutions using different systems, different materials or different specifications. While analyzing, the future costs are brought together into a single value, often by applying the traditional 'present-value' concept on a base date corresponding to the initial costs. The present value can be calculated by the following formula:

$$PV = \frac{C}{(1+r)^t}$$

where
PV = present value
C = cost at current prices
r = selected discount rate per annum expressed in decimal
t = time in years when the future cost is incurred
As an example, if the discount rate is 12% and a sum of \$10,000 is required for maintenance after 5 years, the present value of the sum will be:

$$PV = \frac{10,000}{(1.12)^5} \text{ or about } \$5,675$$

The present values of all maintenance costs to be incurred at intervals of say 5 years up to a specific period (design life of the structure) can be calculated in a similar manner and added to the initial costs to give the net present value (NPV) of a particular system under consideration. Similar values of other options may also be calculated and the results compared for financial evaluation.

Apart from cost considerations, quality of the product or the service provided needs also to be considered while analyzing LCC. Meeting minimum quality standards is an obligatory requirement in all cases.

9.4   FUNDAMENTAL ISSUES

LCC thus involves the following issues:

- Initial costs;
- Future costs;

- Life of the structure;
- Discount rate.

9.4.1 Initial Costs

Initial costs of a structure include costs for acquisition of land, planning, design, construction and testing. The cost of construction includes not only the costs incurred at site, but also the costs incurred in the workshops, transport costs, etc., which are incurred prior to site activities. It is the total of all costs, direct and indirect, in converting a design into a commissioned project ready for operation, i.e., the sum of all costs including labor, supervision, administration tools, raw materials, engineering, installation, commissioning, etc. Investment in a project is met by borrowed capital (thus incurring an interest charge) or by the promoter's own resources (thus foregoing potential interest income). Because of these issues, construction cost should take into account the time value of money in the form of interest on the capital invested for the duration of the project. Thus, a solution with long construction period would attract more interest resulting in higher costs.

9.4.2 Future Costs

Future costs may be either regularly recurring costs or costs which occur occasionally. These costs include the following:

a. *Annual maintenance costs*: these costs comprise regular preventive maintenance costs. The objective of annual maintenance is to carry out the activities so that an acceptable standard of service can be maintained. These include labor, material, equipment, testing costs, structural assessment and design costs incurred in carrying out inspection, preventive maintenance, including minor repair work.

b. *Special inspection and maintenance costs*: most structures need thorough inspection every 5–7 years. The costs include those for inspection and major repair or renewal work and upgrading as necessary.

c. *Renewal of deterioration protection system*: interval of renewal of deterioration protection system depends on the type of the system adopted as well as the environmental conditions. In case of steel structures, prevention of corrosion becomes a very important determinant. Costs involved in carrying out inspection maintenance are to be included under this head also.

d. *Operating costs*: these costs relate to most of the civil engineering structures such as public buildings, airports, marine structures, roads and railways. These include costs for lighting, security arrangement, safety provisions and similar expenses for operating the facilities.

e. *Disruption costs*: the costs due to disruption of the user's normal working during repair or maintenance work are considered as indirect costs by some authorities. This is quite logical as disruptions are bound to affect the user. Thus if a particular area in a factory premises is under repair, production gets disturbed and consequently the net profit of the unit is affected.

Similarly, if a toll road is closed to traffic, not only the earnings from tolls suffer, but also the diversion and consequent delay causes heavy loss to the user. Costs due to such delays need to be considered while assessing the operation costs of an asset.

f. *Occasional costs*: there may be some costs which cannot be predicted at the initial stage of a project, e.g., repair of damage due to accidental impact on buildings, road side structures. There may also be costs for modified demands such as upgrading the runway of an airport or strengthening a structure for heavier loads. Another example of modified demand is widening of a road way or bridge structure to suit specific traffic requirement.

g. *Costs for demolition*: structures that can no longer be used have to be removed. This operation may incur considerable costs. These will include costs for demolition at site, transportation and disposal of surplus materials. The proceeds obtained from sale of scrap material can offset a part of the costs of demolition.

h. *Benefits*: these include all negative costs that accrue to the owner from the structure by way of rents, toll or similar earnings. These are to be credited as and where applicable.

9.4.3   Life of the Structure

For carrying out the costing over the life span of a structure, by implication, the end point of life needs to be defined. There are more ways than one in which this can be done.

Typically, a designer may assume certain specified life span for a structure depending on the material being used. Thus, life of a new structure may be assumed nominally to be 120 years. But long before that the structure may become functionally obsolete due to charge in the technology used for the product for which the structure was initially constructed. Therefore, it may not be appropriate to consider the future costs over the full 120 years. On the other hand, from the investor's point of view, the life of a structure may be assumed to be the time to repay the investment. But actual serviceability of a structure is one matter and repayment of loan is another. In this case, it will also be inappropriate to take the time of repayment of the investment as the life span for computation of future costs. It is thus not easy to arrive at a fool-proof yardstick for assessing the service life of a structure. This would largely depend on the following parameters:

- Quality of material;
- Workmanship;
- Internal environment;
- External environment (including local environments);
- In-use conditions;
- Maintenance conditions.

These parameters may increase or decrease the expected (predicted) service life of a structure, depending on the effect of these parameters on the structure. However, there is need for adequate predictive models and historical data for initiating procedures for using the above parameters.

9.4.4 Discount Rate and Inflation

Discount rate is the rate at which future costs are discounted to the present. It reflects the investor's time value of money. In other words, it takes into account of the fact that money, if not consumed today and invested, grows to be worth more at a future date. Consequently, money received or paid at a future date is not as valuable as the equal amount received or paid today. Thus, discount rate is used to convert costs occurring at different time to equivalent costs at a common point of time. At the same time, it has also to consider that cost increases with the passage of time due to inflation. Inflation is the general reduction in the purchasing power of money from year to year. LCC analysis can be calculated in two ways, viz., including or excluding the effects of inflation. In either case, a consistent system should be adopted throughout the LCC analysis.

Calculation of discount rate, though apparently simple, depends on several uncertain parameters such as inflation, policy of the government and interest rate on borrowings. It is thus not very easy to predict the rate with certainty. However, for comparing different competing solutions, the same discount rate can be adopted for each solution to give sufficiently accurate result for ranking purpose.

9.5 LIMITATIONS

In LCC computation, the parameters that control the results (costs, life of the structure and discount rate) are only assumed values and are, therefore, uncertain. Thus, if these parameters vary, the result will also vary. In order to assess the effects of the variations of the parameters, it is necessary to carry out a number of computations by varying the parameter(s) and arriving at a set of present-value costs. Final selection of the choice is made by sound engineering judgment and other decision criteria.

In the area of maintenance of an asset, LCC approach can be applied both for future assets and for those which are already in existence. However, some perceived difficulties come in the way of practical application of this system in general. Some of these are as follows:

- Useful life of an asset cannot be forecast accurately, particularly when functional, technological or social obsolescence may reach first;
- It is also difficult to forecast the future economic scenario which primarily affects the discount rate;
- It is not easy to predict costs over a long period. Consequently, the future maintenance costs appear to be 'idealistic' as against capital costs which are regarded as 'real';
- As it is, resources are scare. This position imposes natural restrictions on the initial capital expenditure of any project. Therefore, it becomes more difficult to allow extra expenditure towards superior specifications for uncertain future gains by way of lower operating/maintenance costs.

The above difficulties may be real, but are so only varying degrees. No doubt, there are some uncertainties in the parameters used in the method and a certain amount of inaccuracy may creep into the estimates of future costs. However, this should not deter us from using the technique. The results will at least give an idea about the impact of future expenses even if these may not be absolutely correct.

For structures with long lives, the difficulties may perhaps be significant. However, the method is considered more useful for formulating short-term strategies for maintenance, repair and replacement of different elements in buildings and other structures. In such a case, values to be assigned to the different parameters like life expectancy, discount rate and costs can be realistic and the results are likely to be more accurate. Also, because the same parameters are used for different competing solutions, the results can be effectively utilized for comparison purpose.

From the point of view of future costs, one aspect regarding life cycle needs particular consideration. Life expectancy of a structure as a whole may be different from the life expectancy of its individual components. In a factory shed, the life of the structure and that of the cladding will be different. Similarly, in a steel structure, life of paint is much less than that of the steelwork. There may be many more examples like these such as a lift in a building or wearing course in a roadway bridge. Therefore, in order that the life of components can be forecast with fair accuracy, there is need to build up a database indicating life cycles and other relevant information for the critical elements of a structure. Maintenance strategies can be formulated on the basis of the available data by using LCC method.

One impediment for correctly assessing future costs, even for a short time frame, is that while the designer in the project office has all inputs for estimating the initial capital costs, very often, the maintenance costs are not readily available with him, since these are normally maintained by the operation department. There is thus a need to bridge the gap of communication and information exchange between these two units so that a database of historical costs may be set up. This would certainly enhance the reliability of LCC.

9.6   APPLICATIONS OF LCC

Despite certain perceived difficulties, LCC analysis has immense potential in formulating strategies for evaluating different engineering projects. The method brings to focus the fact that future operating and maintenance costs need to be studied and given due importance during the initial design stage itself. Some of the applications of LCC are outlined in the following paragraphs.

9.6.1   Comparing Costs of Alternative Schemes

The costs between alternative schemes can be evaluated by LCC concept taking into consideration future costs and benefits. Thus in an aggressive environment, a special corrosion-resistant painting system which would initially cost more, but would last longer than normal painting system on a steel structure may prove to be more economical over a period of time if analyzed with LCC. Additionally, the life of the structure may be extended, thereby delaying heavy capital expenditure for replacement of the structure.

9.6.2   CHOICE BETWEEN REHABILITATION AND REPLACEMENT

The decision whether to repair just some parts and thereby rehabilitate a structure or to replace the entire structure can be analyzed by LCC. For this purpose, it is necessary first to inspect the structure and evaluate its present condition. If necessary, a rehabilitation scheme has to be designed. The estimated cost of rehabilitation plus future costs with appropriate discounting will have to be compared with the initial cost of the new structure plus discounted future costs before making a suitable decision.

9.6.3   PRIORITIZING OF RESOURCES

Resources are scarce, and it is imperative that these should be used with utmost care. There is thus a compulsive need for prioritizing the allocation of resources for maintenance, rehabilitation and replacement of existing assets. In planning these actions, LCC system can be effectively applied.

9.6.4   BALANCING INITIAL AND FUTURE COSTS

The comparatively recent trend of privatization of infrastructure, particularly in the road development work, has created a new opportunity for the application of LCC concept. Depending on the period of operation after which the project has to be handed over to the authorities, the entrepreneur would like the designer to adopt a solution with comparatively less expensive construction techniques and materials to keep the initial costs low, which may tend to increase the future maintenance costs. The reason is obvious. The initial construction cost has to be financed by the entrepreneur from borrowings carrying high interest rate. It would therefore be expedient for him to keep the cost as low as possible. The higher future costs would be substantially offset by earnings from tolls, advertisements and other such benefits. In such cases, LCC concept can provide useful data for balancing the different options available to the entrepreneur.

9.6.5   IMPROVING PRODUCTIVITY IN THE WORKPLACE

It may be useful to extend the LCC concept to include indirect costs (or benefits) of solutions such as improving internal environment (air conditioning, ventilating, heating, etc.), which will increase the productivity (benefit) of the user over a period.

9.7   STAGES FOR LCC ASSESSMENT

Generally, LCC assessment comprises the following stages:

1. Identification of the target, alternative solutions and constraints for the alternatives;
2. Establishment of the basic assumptions for LCC analysis;
3. Selection of appropriate discount rate and computation of present value of all initial and future costs;

4. Defining the expected life of the asset;
5. Computation of LCC for each alternative solution;
6. Comparison of results and determination of the one with minimum LCC.

9.8   CONCLUDING REMARKS

In spite of its limitations, LCC method has its utilities also. It addresses the future costs in proper perspective and allows for an informed and rational way of comparing alternative solutions over the predicted life of an asset. It gives the designers an awareness of the future consequences of their present actions. It also highlights the importance of good workmanship in workplace in order to reduce the future maintenance costs. For achieving high-quality workmanship, it may be expedient to introduce quality assurance plan (QAP) in the process. LCC method can thus be used as a worthwhile qualitative tool for decision making. However, to make it more effective, the engineer needs reliable inputs such as databases of maintenance costs and service lives as well as appropriate discount rates.

LCC should not be considered as an exact science. It is only an indicative method for studying future costs and giving these costs appropriate weightage in the initial stage of a project.

BIBLIOGRAPHY

Ghosh, U.K., 1997, *Life Cycle Costing*, Paper presented in the Institution of Engineers, Kolkata.
Jones, A.E.K. and Cussens, A.R., 1997, *Whole-life Costing*, The Structural Engineer, April 1997, Institution of Structural Engineers, London.
Mott, G., 1992, *Investment Appraisal*, Pitman Publishing, London.
Standard Practice for Measuring Life-Cycle Costs of Buildings and Building Systems, ASTM: E 917–94, American Society for Testing and Materials, Philadelphia, PA.

10 SWOT Analysis

10.1 INTRODUCTION

Currently, most organizations engage strategic planning techniques for helping them to monitor allocation of the resources in the best possible way to achieve optimum productivity. SWOT analysis is such a strategic planning technique.

It is generally used at the initial stage of a project to identify the areas where the project can do well, or where it may need improvement. The technique is also used retrospectively for identifying the positive and negative aspects of an ongoing project *vis-à-vis* what has been achieved in reality. SWOT is an acronym for the following four parameters:

- **S**trengths: attributes of a project that give it an advantage over other projects.
- **W**eaknesses: attributes of the project that place it at a disadvantage relative to other projects.
- **O**pportunities: elements in the surrounding setting that the project can exploit to its advantage.
- **T**hreats: elements in the surrounding setting that could be harmful for the project.

Strength and weaknesses relate to internal factor within the organization and its products or services, whereas opportunities and threats relate to external factors over which the organization has little control. SWOT analysis helps the stakeholders to understand the internal strengths and weaknesses of a project or business as well as to identify the opportunities and threats in the marketplace.

SWOT analysis emerged as an aid for strategic planning in the 1950s and has been a very popular technique since then. The technique is primarily used for acquiring information, based on which the strategy can be developed.

10.2 TEAM EFFORT

SWOT analysis requires a team effort. A group of analysts should participate in the discussions or brainstorming sessions. It is necessary that the analysts should have an insight about the organization and its business environment. Preferably, their expertise and backgrounds should be different.

On completion of SWOT analysis, the team's recommendations should be effectively utilized in the strategic planning process.

10.3 PRESENTATION OF SWOT ANALYSIS

For undertaking SWOT analysis of a project, the strengths, weaknesses, opportunities and threats are normally listed on the same page. The page is segmented into

TABLE 10.1
SWOT Analysis format

	Positive/Helpful Aspects	Negative/Harmful Aspects
Internal Factors	Strengths	Weaknesses
External Factors	Opportunities	Threats

four rectangles, and strengths and weaknesses are entered in the top rectangles and opportunities and threats in the bottom rectangles. Table 10.1 illustrates a SWOT analysis format with four elements in a 2×2 matrix. It may be noted that internal factors are listed in the top row, while external factors are listed in the bottom row. Also, the positive (helpful) aspects are in the first column, while the negative (harmful) aspects are in the second column.

10.4   INTERNAL FACTORS: STRENGTHS AND WEAKNESSES

Strength relates to a productive and advantageous attribute that adds value to something and makes it comparatively more exclusive than something else. Thus, strength is a positive, favorable and advantageous characteristic. At the organizational level, strength involves ability by which an organization gains advantage over its competitors. Weakness relates to a condition in which the competence necessary for accomplishing something is not enough or totally absent. Weakness is thus a negative and unfavorable characteristic. At the organizational level, weakness implies inability of the organization to overcome its deficiencies. Thus, organizational weakness comprises the aspects or activities in which the organization is less capable and less proficient in comparison with its competitors. These are negative aspects that weaken the performance of the organization, and lead to its inability to respond to the possible problems or opportunities, and to adapt to changes. It is important for an organization to be aware of its weaknesses, as these have the potential to lead it to incompetence and futility.

Areas that are normally identified as strengths and weaknesses are outlined below:

   a. *Human resource:* this is the fountainhead of strength for any organization. It comprises the staff, the managers, board members as well as target population. Their knowledge, expertise, experience, etc. can be used to the organization's advantages. Thus, the human resources development policy is very important for the organization.
   b. *Physical infrastructure:* the physical infrastructure, such as the land and its location, building, machinery and equipment, is an area of strength or weakness of the organization. If the size of the infrastructure is large, the organization may be able to venture into certain projects which its competitors would not even think of. The converse is also true.
   c. *Financial resources:* availability of finance is an important feature of strength for any organization and is an advantage over its competitors. It provides opportunities to expand the organization's activities by way of

entering into new markets and new customers as well as launching new products in these areas.

A strong cash flow is also a potential strength from financial perspective. On the contrary, lack of cash may result in business failure. Strong cash flow also allows the organization to provide extended credit to its customers and better terms to its suppliers. These could be considered as competitive strengths.

d. *Distinctive expertise:* the organization may have as its strength, some distinctive expertise on a particular process or activity. Such expertise is often defined as the core competence of the organization. Examples are departmental hierarchies and software systems employed.

e. *Image:* the other source of competitive advantage is the image of the organization or the product in the marketplace, i.e., what the customer believes to be useful. A powerful image is important for any organization or product for negotiating with other organization for partnerships or entering into a fresh market. The goodwill created by the image may also attract high-quality human resources to the organization.

f. *Client experience:* clients' experience with an organization encompasses its different sections, such as sales, marketing, delivery, customer service, after-sales service and accounts. These areas provide chances of interaction between the clients and the organization. Positive interaction creates a positive reputation in the marketplace and enhances customer satisfaction leading to profitability.

10.5   EXTERNAL FACTORS: OPPORTUNITIES AND THREATS

Development of strategic planning process depends largely on certain factors which are external to the organization. These external issues will also affect an organization in the same competitive marketplace as the internal factors discussed in the earlier section. Analysis of the external environment will relate to the opportunities and threats that the planning process has to go through. These are briefly outlined below:

a. *Political factor: Government policies*: government policies on various issues, such as taxation and employment legislation, have considerable impact on the working of an organization. Increasing globalization of international markets has enhanced harmonization between countries. Consequently, political change or unrest in one country may have far-reaching effect on other countries too. It becomes imperative for organizations of one country to develop their strategy in compliance with the political developments in other countries also.

b. *Economic scenario*: direction of strategic policies of any organization depends largely on the domestic economy of the country. Domestic economy influences the customer demand. With recent trend of globalization of world markets, however, a country's domestic economic scenario can change with the change in the economic scenario of another country. It is therefore becoming necessary for organizations to develop flexibility in their strategy in order to respond to external situations on which they do not have direct control.

c. *Changes in technology*: change in technology is an ongoing process. In recent decades, impact of technological innovation as well as emergence of new technologies has led to far-reaching changes in most organizations. Of late, improvements in technologies have become an instrument of radical change, to the advantage of the customers as well as to internal operations of the organizations. In order to respond to the unpredictable advancement in technology, it has become important for organizations to be more flexible in order to seize any opportunity for future development.

d. *Social and cultural environment*: changes in social and cultural environment of a society occur rather gradually and take considerable time. Nevertheless, these changes influence an organization's strategy all the same. Technological advancement during the last decades of the 20th century, in particular in the field of digital technology, has considerable impact on the social and cultural environment. Many organizations have taken advantage of this advancement in projecting their products to prospective customers via social networks across the digital media.

e. *Competition*: characteristic features and ambit in the market environment have distinct impact on the strategic direction of an organization. It is therefore imperative that an organization should analyze its position in this respect and approach appropriately its prospective customers in order to wean them off from the competitors.

10.6   SWOT ANALYSIS PROCESS

SWOT analysis can be conducted as an individual exercise. However, better results can be achieved if this is carried out by a group of analysts whose backgrounds differ. It is best conducted in a group in a brainstorming session in which team members are responsible for decision making and strategic planning of the organization. In such a case, participants can offer different perspectives based on their different experiences on the strengths and weaknesses of the process. Likewise, participants may also report about opportunities and threats, which are also important for developing the strategy.

In practice, SWOT analysis process is the same irrespective of whether this is done for future planning of specific products, work or any other area. The various steps for the analysis are as follows:

- Step 1: Constitute a group of analysts with diverse expertise and backgrounds.
- Step 2: Identify and list all strengths, weaknesses, opportunities and threats related to the alternative decisions being examined.
- Step 3: Summarize the results for each alternative in a four-cell SWOT format.

These results need to be reviewed based on which strategy for action is taken. A typical example of SWOT analysis for a product is shown in Table 10.2.

TABLE 10.2
Typical example of SWOT Analysis

Strengths	Weaknesses
Satisfactory quality	High price
Robust in construction	Lack of fund
Variety of sizes available	
Product designed by specialists in the field	
Opportunities	**Threat**
Good prospect in emerging global market	Negligible growth in traditional market segment
Development of other products	Products with cheaper raw materials by competitors

10.7 REVIEWING SWOT ANALYSIS RESULTS

SWOT analysis results will provide a clear image of the external as well as internal issues which have considerable influence on the organization's future. Based on these results, the strategy for action needs to be chosen in order to reduce organizational weaknesses and guard against external threats. Additionally, external opportunities are to be exploited effectively. The directional output of SWOT analysis will have immense influence on the strategy of the organization in different issues.

10.8 ADVANTAGES OF SWOT ANALYSIS

SWOT analysis is a popular model for making decisions by individuals or organizations. Characteristics that can be considered as advantages include the following:

- SWOT analysis has a general perspective and thus presents general and broad solutions. It is a technique that satisfactorily performs macro-evaluations;
- SWOT focuses on both positive and negative aspects of the internal and external settings of an organization;
- SWOT analysis method is simple and straightforward and inexpensive;
- SWOT analysis helps organizations to discover the external opportunities to their advantage and eliminates the internal threats;
- SWOT analysis by an organization's competitors can help in formulating strategies to the advantage of the analyzing organizations (competitors);
- SWOT analysis is normally conducted in group discussion (brainstorming sessions) and promotes pool of knowledge;
- SWOT analysis forms a thinking model comprising information collection and its interpretation for arriving at strategic decisions;
- SWOT analysis encourages management to adopt a futuristic approach in preference to current situation and problems;
- SWOT analysis enjoys a broad perspective and can be applied at various levels, e.g., individual level, organizational level, national level as well as international level. It can be usefully utilized by various organizations, governments, project authorities, etc. to their advantage.

10.9   LIMITATIONS OF SWOT ANALYSIS

Although SWOT analysis is one of the most widely used techniques for strategic management process, it has its limitations too. These limitations are briefly discussed in the following paragraphs:

- SWOT analysis deals primarily with current scenario in respect of strengths, weaknesses, opportunities and threats. Therefore, it needs to be re-done for ascertaining the strategy for a period other than the present;
- An item can be strength for one issue, but may be threat for another issue inside the same organization. Also, opportunity not availed by one organization may be taken up by its competitors and becomes a threat to the former. Additionally, if an organization is obsessed with a single strength, such as cost control, it may ignore its other weaknesses, such as in quality assurance plan (QAP) and human relationship development policy. Thus, classification and posting of variables into one of the four quadrants of SWOT analysis format is not a fool-proof operation. It may lead to misrepresentation of the internal and external issues;
- Compared to the benefits, the SWOT procedure by assembling a group of experts and carrying out brainstorming sessions is time-consuming for participating officials. It has thus high cost, but fewer benefits;
- SWOT analysis does not provide any quantitative guideline regarding the size of the competitive gaps between an organization and its competitors.

The above-mentioned limitations indicate that SWOT analysis provides only a direction for further in-depth strategic analysis. It is thus not an end in itself and only raises awareness about important issues.

10.10   APPLICATIONS OF SWOT ANALYSIS

SWOT analysis may be applied in any decision-making situation in order to achieve a desired objective. It can be used in a government department, a nongovernment organization or even in individual setups.

- SWOT analysis can be used to examine the currently prevailing scenario of the operations in an organization – whether it is performing satisfactorily or needs improvement;
- Timely SWOT analysis can show the ways for better utilization of the strengths of the organization and improve upon the weaknesses. Thus, it is useful for making strategic decisions for future planning or preventive crisis management;
- SWOT analysis is useful in prioritizing the actions for any improvement that may be required in an organization;
- SWOT analysis can also be used for personal development, career improvement or an individual's problem solving;

- It is often necessary to analyze the competitor's strength, weaknesses, etc., in order to ascertain their relative competitive status. It is useful to focus on the different aspects of the competitor's cost structures, resources, experiences, expertise, etc. It will be useful to collect information about these aspects.

10.11 CONCLUDING REMARKS

SWOT analysis provides a panoramic view of the strengths, weaknesses, opportunities and threats in respect of a concept or endeavors that an organization or an individual intends to promote or undertake. It is a very useful method for planning and decision making. While a number of other methods of analysis do exist for attaining long-term objectives in an organization, SWOT method has been very popular. It is a simple method for analyzing internal and external surroundings for making strategic decisions in respect of an industry, an organization, an individual, a project, a product, etc.

BIBLIOGRAPHY

Gould, R., 2012, *Creating the Strategy*, Kogan Page Ltd., London.
Samset, K., 2010, *Early Project Appraisal*, Palgrave Macmillan, London.
Westwood, J., 2013, *How to Write a Marketing Plan*, Kogan Page Ltd., London.

Index

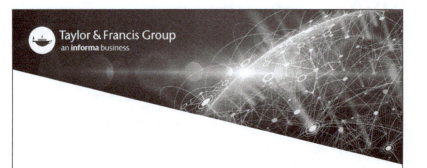